NATO ASI Series

Advanced Science Institutes Series

A Series presenting the results of activities sponsored by the NATO Science Committee, which aims at the dissemination of advanced scientific and technological knowledge, with a view to strengthening links between scientific communities.

The Series is published by an international board of publishers in conjunction with the NATO Scientific Affairs Division

A	Life Sciences	Plenum Publishing Corporation
B	Physics	London and New York
C	Mathematical and Physical Sciences	D. Reidel Publishing Company Dordrecht and Boston
D	Behavioural and Social Sciences	Martinus Nijhoff Publishers Dordrecht/Boston/Lancaster
E	Applied Sciences	
F	Computer and Systems Sciences	Springer-Verlag Berlin/Heidelberg/New York
G	Ecological Sciences	

Series E: Applied Sciences – No. 95

Problems in Service Life Prediction of Building and Construction Materials

Edited by

Larry W. Masters

Organic Building Materials Group
Center for Building Technology
National Bureau of Standards
Gaithersburg, MD 20899
USA

1985 **Martinus Nijhoff Publishers**
Dordrecht / Boston / Lancaster
Published in cooperation with NATO Scientific Affairs Division

Proceedings of the NATO Advanced Research Workshop on Problems in Service Life Prediction of Building and Construction Materials, Paris, France, September 10-12, 1984

NATO Advanced Research Workshop on "Problems in
Service Life Prediction of Building and Construction
Materials" (1984 : Paris, France)
Problems in service life prediction of building and
construction materials.

(NATO ASI series. Series E, Applied sciences; no.
95)
"Proceedings of the NATO Advanced Research Work-
shop on 'Problems in Service Life Prediction of
Building and Construction Materials', Paris,
France, September 10-12, 1984"--T.p. verso.
1. Building materials--Service life--Congresses.
I. Masters, Larry W. II. North Atlantic Treaty
Organization. III. Title. IV. Series.
TA409.2.N37 1984 691 85-10577
ISBN 90-247-3181-X

ISBN 90-247-3181-X (this volume)
ISBN 90-247-2689-1 (series)

Distributors for the United States and Canada: Kluwer Boston, Inc., 190 Old Derby Street, Hingham, MA 02043, USA

Distributors for the UK and Ireland: Kluwer Academic Publishers, MTP Press Ltd, Falcon House, Queen Square, Lancaster LA1 1RN, UK

Distributors for all other countries: Kluwer Academic Publishers Group, Distribution Center, P.O. Box 322, 3300 AH Dordrecht, The Netherlands

Printed in The Netherlands

LIST OF ORGANIZING COMMITTEE MEMBERS

Larry W. Masters, Chairman
National Bureau of Standards
Gaithersburg, MD USA

Phillippe Eurin
Centre Scientifique et Technique du Batiment
Paris, France

Pieter C. Kreijger
Technische Hogeschool
Eindhoven, The Netherlands

Peter L. Pratt
Imperial College of Science and Technology
London, England

Gerhart I. Schuëller
Universität Innsbruck
Innsbruck, Austria

Christer Sjöström
The National Swedish Institute for Building Research
Gävle, Sweden

PREFACE

Degradation, the chemical/physical response of building and construction materials exposed to in-service environments, must be predicted prior to their installation in structures if materials are to be effectively selected, used and maintained. These assessments of materials degradation require that methods be available to aid prediction of service life.

The objectives of building materials science are a) to characterize and categorize materials, b) to predict, preferably in a mathematical sense, material or component response including expected service life, and c) to make improvements in material response through improvements in design, formulation, processing or specification.

For building and construction materials, continued progress has been made towards objective (a), but little progress has been made towards objectives (b) and (c). Of these, the mathematical prediction of service life appears to be of greater importance, because, if general approaches or models having application to a wide range of building and construction materials can be identified, then the categorization, selection, use and improvement of materials can proceed in a systematic manner.

Researchers in advanced technologies, such as aerospace, nuclear, electronics and medicine, have apparently been more successful than researchers in building and construction technology in responding to the need for reliable predictions of service life. For these advanced technologies, statements about the expected or minimum life expectancy of the materials, components or systems under the intended operating conditions are required prior to installation. These statements are often probabilistic in nature and are based upon reliability theory and life testing analysis techniques. It is evident that the operating conditions for many advanced technology materials are comparable to those to which building and construction materials are exposed and that failure mechanisms are often similar. This, together with the fact that the theory used for predicting service life in advanced technology materials is material independent, implies that a direct transfer of knowledge to all materials is possible. However, as yet, the building community as a whole has little knowledge of these analytical procedures.

It is with the above background information that a NATO Advanced Research Workshop on Problems in Service Life Prediction of Building and Construction Materials was held September 10-13, 1984 at the Fédération Nationale du Batiment, Saint Rémy-lès-Chevreuse, France. The objective of the Workshop, which was attended by 32 researchers from 16 countries, was to explore opportunities for applying, to research on building and construction materials, analytical procedures which have been utilized in predicting service life of materials, components and systems in advanced technologies. To pursue this objective, the Workshop brought together researchers from selected advanced technologies and researchers from building and construction technology. The Workshop included six sessions:

1. State-of-the-Art of Service Life Prediction of Building and Construction Materials,
2. Approaches to Service Life Prediction in Advanced Technologies,
3. Commonalities of Problems,
4. Mathematical Analysis Techniques Used in Advanced Technologies,
5. Mathematical Analysis Techniques Used in Building and Construction Technology, and
6. Formulation of Recommendations.

It included twelve invited lectures, small group discussions, reports by group discussion leaders, open discussion periods, question and answer periods with lecturers and a closing summary.

This text presents the Proceedings of the Workshop.

ACKNOWLEDGEMENT

The editor acknowledges, with sincere appreciation and with a sense of humility, the extensive contributions made to the Workshop and to these Proceedings by Mrs. Beverly J. Hale, Secretary to the Organic Building Materials Group of the National Bureau of Standards. Mrs. Hale spent countless hours, frequently beyond her normal working day, handling the extensive details of meeting arrangements, program printing, mailings, financial matters and numerous other activities associated with the Workshop and its Proceedings. That the editor received numerous compliments from participants for excellent organization and attention to details is, in large part, due to the outstanding contributions of Mrs. Hale.

The editor also takes this opportunity to express sincere appreciation to organizations and individuals listed below for their contributions to the success of the Workshop:

- The NATO Scientific Affairs Division, Brussels, Belgium for providing the funding needed to hold the Workshop.

- Dr. Robert Carnahan, Vice President of Science and Technology, United States Gypsum Company, Chicago, Illinois for suggesting that a NATO Workshop be held and for assisting in obtaining the NATO funding.

- Dr. Geoffrey Frohnsdorff, National Bureau of Standards, Gaithersburg, MD, for submitting the proposal to NATO.

- Dr. J. Bresson and the staff at the Fédération Nationale du Batiment for providing outstanding meeting facilities and outstanding assistance during the Workshop.

- Ms. Denise L. Herbert, National Bureau of Standards, Gaithersburg, MD, for typing extensive portions of these Proceedings.

- The Public Information Division, National Bureau of Standards, for assistance in printing of the Workshop Program and in handling other details of the meeting.

- Participants at the Workshop for providing stimulating ideas and discussion; lecturers for providing excellent papers; and discussion leaders for effectively leading and reporting small group activities.

- Organizing Committee members (Philippe Eurin, Pieter C. Kreijger, Peter L. Pratt, Gerhart I. Schuëller and Christer Sjöström) for excellent assistance in planning and implementing the Workshop.

- Martinus Nijhoff Publishers and, particularly, Mrs. Henny Hoogervorst for outstanding responsiveness in producing this text.

TABLE OF CONTENTS

I. STATE OF THE ART OF SERVICE LIFE PREDICTION OF BUILDING AND CONSTRUCTION MATERIALS

OVERVIEW OF METHODOLOGIES FOR PREDICTION OF SERVICE LIFE

Christer Sjöström

Head of Division, Materials and Structures
National Swedish Institute for Building Research
Box 785, S-801 29 Gävle, Sweden

ABSTRACT

Most research institutes and laboratories engaged in studies of the service life of building materials follow some established procedure when performing the investigations. There are also examples of systematic approaches proposed by international organizations and some methodologies are established as national standards. The paper presents an overview of some proposed or used methodologies for approaching research on the service life of building materials and components. A joint committee of the international organizations CIB and RILEM is preparing a generic methodology which is meant to meet the need of the two organizations in their research on service life prediction. This work is also accounted for and the article ends with a review of some of the general research needs in connection with the prediction of service life.

1 INTRODUCTION

The prediction of service life of building and construction materials is a true example of a multi-disciplinary field of work. Knowledge gained from materials science, laboratory testing and in-service studies is necessary in a prediction and the forecast must be based upon a proper characterization of the expected service conditions. A complete prediction ought to contain the creation of a mathematical model to document the observed degradation and to enable the calculation of the acceleration factor. Service life prediction involves considerable judgment and numerous assumptions that have to be accounted for and sometimes explained.

In such a complex research area, systematic approaches or methodologies have an important function. They act as checklists when an investigation is performed, ensuring that vital steps of the work are not forgotten and giving guidance on how to design and to operate the testing and on the interpretation and presentation of data. An agreed systematic methodology for approaching research on service life prediction would help in communication between those active in the field, aid in linking various elements of research and support the identification of research needs.

The objectives of this paper are to give an overview of some proposed and used methodologies and to point out the common parts of the approaches. The technical barriers to reliable predictions are many and how these obstacles are met by the different methodologies is reviewed. A main source of information for the paper is the work being done by the joint committee on prediction of service life that has been established by CIB 1) and RILEM 2). The work of the committee is so far documented in a draft report (1). The committee is, among other undertakings, preparing a generic methodology which is also presented in this article.

In section 6, the general research needs that are connected with the prediction of service life are pointed out.

2 TERMINOLOGY

The question of building materials durability meets a growing interest in most countries and there is also increasing international cooperation in the field of durability research. A main difficulty in the communication between researchers is the lack of uniform terminology. The different interpretations of terms such as durability and performance over time are for example reviewed by Sneck in (2) and (3).

In the terminology adopted by the joint CIB/RILEM committee, durability is viewed as the resistance of a material, incorporated in a structure, to the environmental factors or agents. The

---

1. International Council for Building Research Studies and Documentation
2. International Union of Testing and Research Laboratories for Materials and Structures

surrounding materials are a part of the environment and they influence, together with the design of the structure as a whole, the service life of the material in question. With this view, durability means the resistance of a material to deterioration.

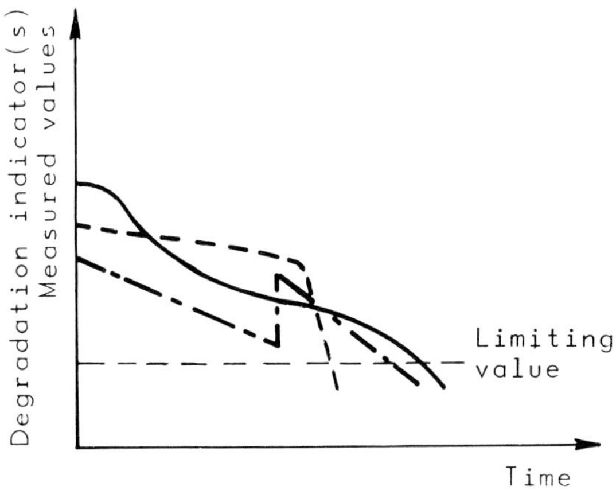

Figure 1 Hypothetical performance over time functions

The extent of deterioration can only be assessed when the specific measurable properties have been chosen which can be used as indicators of degradation.

Performance over time can be seen as the function describing how the measured values of the chosen properties vary with time. With this function established and with the limiting acceptable values of the indicators defined, all essentials are known for a service life prediction. It can be stated that the evaluation of service life aims at the determination or prediction of the performance over time functions.

3 WHAT ARE THE GENERAL BARRIERS TO PREDICTION OF SERVICE LIFE?

The two main sources of service life data are experience and testing. For materials have been used in buildings for a long period of time, a prediction of service life might proceed from in-service performance data. But experience is not always a reliable source of knowledge. It is critical how the data are obtained and reported and this will be further commented on in Section 6, General Research Needs

For new materials and for all instances where materials are subjected to new conditions, testing is the remaining alternative for generating service life data. It is well known that, despite the many existing test methods for assessing the relative durabilities of building materials, the data obtained from these methods are seldom adequate for reliably predicting service life. The need for improved test methods is a main technical impediment to the prediction of service life.

Masters (4) has pointed out a number of barriers to service life prediction and to the development of improved test methods. Among the needs, he mentions the demand for: 1) a systematic approach or methodology for treating the problem, 2) an effective mechanism for obtaining and reporting data on the actual in-service performance of materials, 3) knowledge of the mechanisms of degradation, 4) knowledge of the environmental factors causing degradation, 5) the ability to simulate or account for the synergism between degradation factors, and 6) mathematical models describing the material behavior in specific environments or applications.

In sections 4 and 5, the question of systematic approaches for treating service life prediction is presented in detail. The commonalities of some examples are reviewed, as well as how the different methodologies complement each other.

4 SOME EXAMPLES OF SYSTEMATIC APPROACHES

Most research institutes and laboratories engaged in studies of building materials durability follow some form of routine or guideline when performing their investigations. These guidelines have practical purposes, such as ensuring that the studies are carried out in a logically, technically and economically sound manner using the appropriate facilities with regard to the material or problem under investigation and making certain that the results are presented in a uniform and easily understandable way.

There are also several examples of systematic approaches proposed by international organizations and some methodologies are established as national standards.

The examples presented below represent all of the above categories. The scopes of the approaches are different and more or less delimited and, consequently, the lay-out varies.

4.1 Methodology for Developing Accelerated Tests, ASTM E632

To provide a more general approach to durability testing, a committee of the American Society for Testing and Materials (ASTM) has presented the ASTM Standard Practice for Developing Accelerated Tests to Aid Prediction of the Service Life of Building Components and Materials (5). This established framework has acted as a model for several methodologies developed and adopted by specific laboratories.

Figure 2 shows a chart indicating the steps that are recommended to be undertaken when developing and applying tests for predicting th service lives of building materials. The framework is divided into four parts: Problem Definition, Pre-testing, Testing and Interpretation and Reporting of Data. Of special importance in the practice are the parts pointing out the necessity of identifying the critical performance characteristics that can serve as indicators of degradation, the expected type and range of degradation factors and the possible degradation mechanisms. According to the practice, when the predictive service life test has been designed and the test is performed, the results should be compared with the findings from a long-term test under service conditions.

Only if that comparison shows that the results from the two test situations are consistent with each other, can the next part, Interpretation and Reporting, be undertaken. This part also includes the vital step of documenting by mathematical models the direction of degradation.

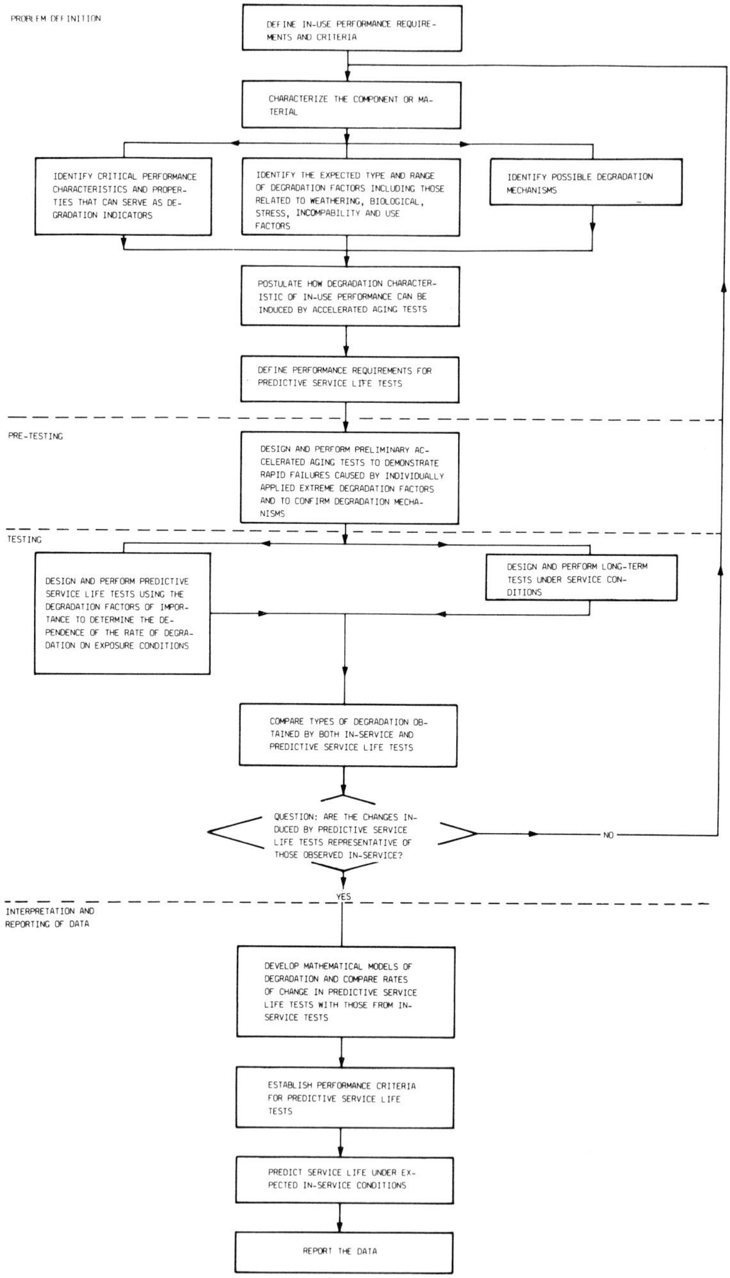

Figure 2. Procedure for service life prediction as outlined in ASTM E632

4.2 Examples of Methodologies Derived from ASTM E632

Two very similar examples of procedures that are slight modifications and simplifications of ASTM E632 are the ones used by Centre Scientifique et Technique du Batiment (CSTB) and the National Swedish Testing Institute. The latter approach is shown in figure 3.

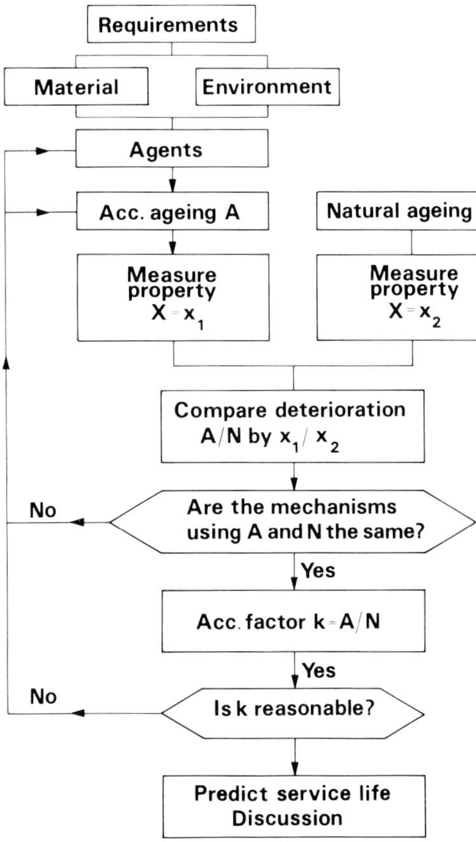

Figure 3. Procedure for service life prediction used by the National Swedish Testing Institute

4.3 Methodology Considered by RILEM TC-60 CSC on Corrosion of Steel in Concrete

Figure 4 shows a methodology proposed and discussed within the RILEM Committee TC-60 CSC and within Comité Euro-International du Béton (CEB). The approach points out the need for defining the durability requirements and for knowing the deterioration processes. It is suggested that the materials behavior be formulated by mathematical models and that the essential and also the uncertain parameters contributing to the deterioration be defined. The methodology calls for the elaboration of simplified models and for the application of probabilistic methods.

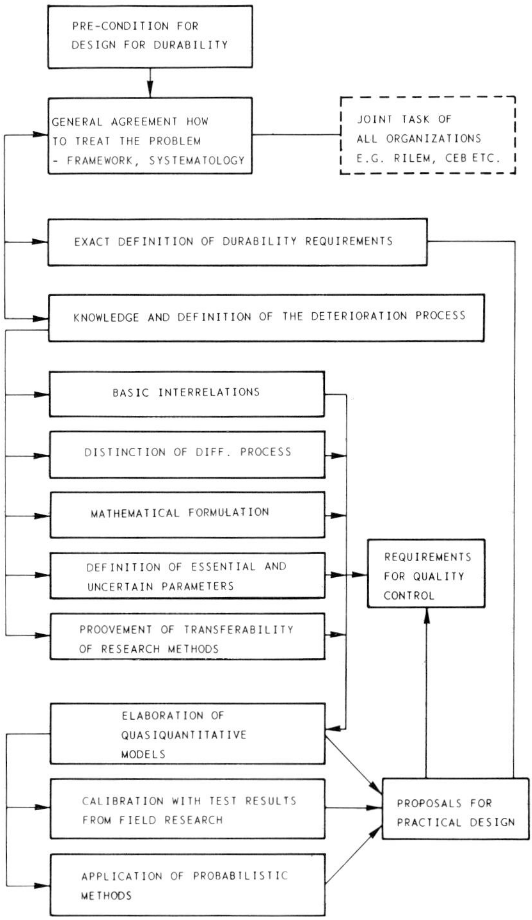

Figure 4. Methodology discussed within RILEM TC-60 CSC

4.4 Methodology for Studies of Solar Photovoltaic Modules

The Jet Propulsion Laboratory is using the methodology shown in figure 5 for service life prediction of solar photovoltaic modules (6). The analysis of the degradation processes is performed in six steps:

1. Identification of the loads from the environment and from the application
2. Description of the response of each material or component to the loads
3. Determination of the material changes, i.e. the nonreversible response to loads
4. Consideration of the damage mechanism, i.e. the loss of integrity or decrease in performance
5. Determination of the failure mode (failure is defined as a performance loss large enough to require repair or replacement)
6. Assessment of the performance penalty.

Though this framework is developed for a special field of technology, the general build-up of the approach is also applicable to other areas.

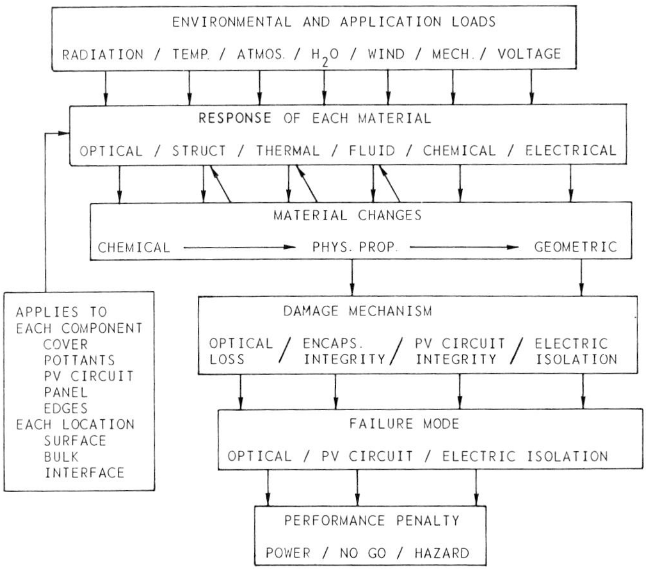

Figure 5. PV module failure - analysis matrix used by Jet Propulsion Laboratory

4.5 Methodology of Australian Standard 1745, Part 2, 1975

This example of a national standard (7) for the prediction of service life is concerned with the natural weathering of plastic materials. As can be seen in figure 6, the procedure contains four main parts: Scope, Definitions, Guidelines for preparation of data sheets and Guidelines for the use of the data sheets. Regarding the outdoor weathering part of the service life analysis, the standard refers to a special Australian standard (8).

1. SCOPE

2. DEFINITIONS
   - 2.1 Unit manufacturing process
   - 2.2 Grade - the material used in the unit man proc
   - 2.3 Solar exposure period (recorded in h of sunshine)
   - 2.4 Total exposure period (recorded in months or days)
   - 2.5 Failure - when the conditions of the article is no longer acceptable

3. GUIDELINES FOR PREPARATION OF DATA SHEETS
   - 3.1 General - possible manufacturing processes, applicat etc
   - 3.2 Guidelines for the selection of suitable tests
   - 3.3 Samples for exposure
   - 3.4 Climate data to be recorded during tests
   - 3.5 Exposure of samples
   - 3.6 Presentation of data for design and selection purposes
     - a. material
     - b. unit manufacturing process
     - c. application envisaged for grade
     - d. changes in properties with exposure
     - e. exposure data
     - f. location of test site
     - g. climatic data

4. GUIDELINES FOR THE USE OF DATA SHEETS
   - 4.1 General
   - 4.2 Use factor
   - 4.3 Anticipated modes and criteria of failure
   - 4.4 The relationship of failure with exposure test data
   - 4.5 Geographic adjustment of weathering results

Figure 6. Procedure for predicting service life as recommended in AS 1745

4.6 Approach Discussed by CIB Working Commission W60

The CIB Working Commission W60 on the Performance Concept in Building is discussing an approach to make the concept of durability manageable. According to the approach, durability can be subdivided into four main manageable aspects, Materials, Design, In-Use Conditions and Maintenance, as can be seen in figure 7. Each of these items can be further subdivided and that detailing is most often necessary. Blach and Brandt present (9) a number of examples of possible subdivisions. Each property is evaluated using banded levels and a combination of the banded scales creates a performance or durability profile, as illustrated in figure 8.

Unlike most of the other methodologies presented, this is not an approach for a step-wise service life prediction but a way of presenting the data that are vital for a selection of a material or a technical solution.

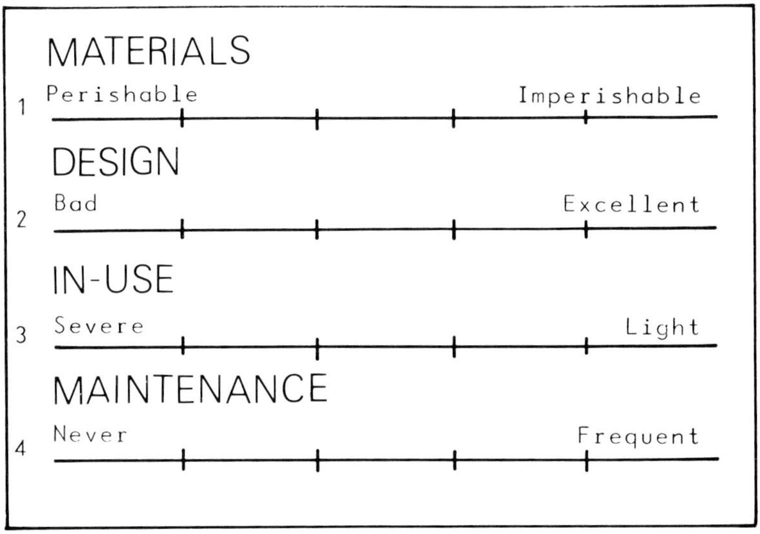

Figure 7. Durability sub-divided into four manageable aspects

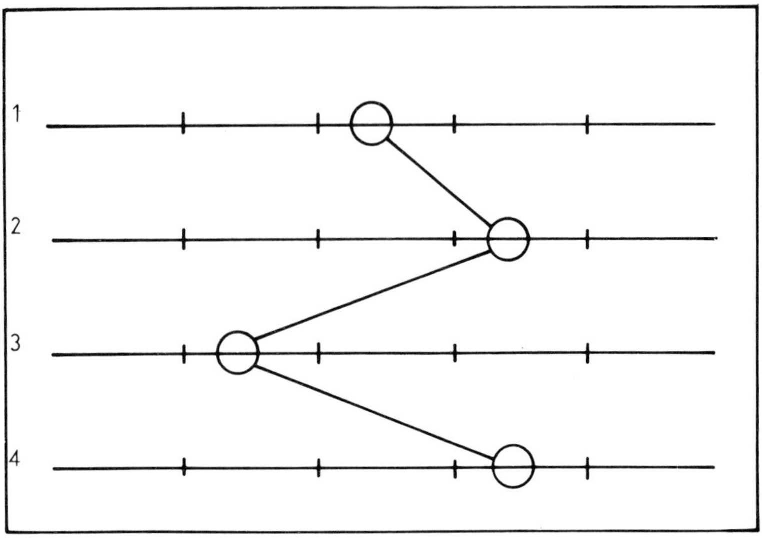

Figure 8. Example of a durability profile

4.7 Common Parts of the Different Approaches

In concentrating on the methodologies that are explicitly concerned with the prediction of service life, it is easy to find similarities between the approaches, though the lay-outs are different. The importance of identifying the degradation mechanisms is emphasized as the need for comparing the results from predictive service life tests or artificial aging with the findings from long term tests under service conditions. The determination of the performance requirements and of the materials performance characteristics, and the identification of the vital degradation factors are other common parts.

Some of the methodologies clearly point out the development of mathematical models as an important step of a service life prediction. The procedure discussed by RILEM TC-60 CSC takes an interesting further step by arguing for the application of probabilistic methods.

Looking at the barriers to service life prediction described by Masters (4), most of the methodologies at least implicitly treat these obstacles.

The special approach for reporting the data that determine the service life presented by CIB W60 can serve as a useful complement to the other methodologies.

5 A GENERIC METHODOLOGY

The overview of approaches not only demonstrates several similarities, but the different activities also clearly show the need of such methodologies. An internationally accepted methodology for treating the problem of service life prediction would lead to many advantages, by making the communication and cooperation between researchers easier.

The draft report of the CIB W80/RILEM 71-PSL Committee (1) lists the essential elements of a systematic methodology that also are a prerequisite to make a methodology widely adopted. Presented in brief, the elements are:

1. It must be generic, meaning that the methodology must be applicable to a broad range of building materials and components.

2. It must lead to identification of the data needed, such as data on degradation factors, degradation mechanisms, performance requirements, intended maintenance methods, design features etc.

3. It must be based upon reliable test methods. Good correlation between laboratory and in-service behavior of the material should exist. The test result should be repeatable and reproducible.

4. It must provide guidance on interpretation of data, meaning that it must aid in predicting service life from the available data. Suitable methods for this purpose, e.g., mathematical models, must be listed.

5. It must lead to documentation of assumptions made.

Based upon these elements and upon a review of the approaches presented in section 4, the Committee is discussing the methodology outlined below.

5.1 Systematic Methodology Outlined by CIB W80/RILEM 71-PSL

Figure 9 shows the five main parts of the methodology: Definition, Preparation, Pretesting, Testing, and Interpretation and Discussion.

The problem definition step covers an analysis of the problem under study and an identification of essential data, such as the user needs, the building context, the performance requirements and performance criteria and a characterization of the material or component.

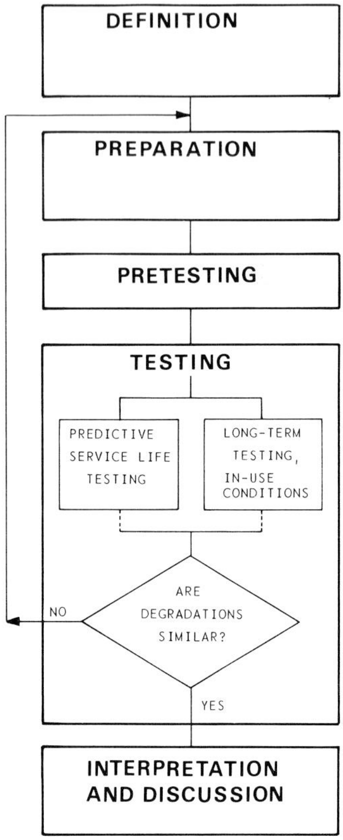

Figure 9. Methodology outlined by CIB W80/RILEM 71-PSL

The preparation step contains the identification of degradation factors, possible degradation mechanisms and a postulation of how degradation can be accelerated by aging tests. The material characteristics or properties that can be used as indicators of degradation shall be identified.

The purpose of the pretesting is to show how rapid changes of the material can be induced by extreme levels of the degradation factors and to check the previously identified degradation mechanisms. The pretesting may also lead to a better understand ing of the primary degradation factors and help in the choice of degradation indicators. The results of the pretests should be used to adjust, if necessary, the previous assumptions.

The purpose of the testing is to create the necessary data for a prediction of service life. It appropriate methods are not available, new or improved predictive service life tests must be designed. Important steps are to determine the relationships between the rates of degradation and the exposure conditions, and to perform tests under in-use conditions to confirm that the degradation mechanisms induced by accelerated tests are correct. The predictive tests should normally be designed from the information gained in pretests. Long term aging tests may be studies of the complete system under in-service conditions or exposure of the selected materials at outdoor weathering sites. The types and range of degradation obtained in the accelerated aging test should be compared with the observations from the long term test under in-use conditions. Only if the degradations are similar can the findings be used for a prediction; if not, the information from the Preparation and Pretesting steps must be taken under new consideration.

Interpretation and Discussion covers the reporting of the data obtained in testing, and prediction of the service life of the investigated material or component or a comparison of the relative durabilities of the studied products. Judgment by experts is always an important part of the interpretation of data and must be accounted for. All assumptions must be described. The relation between the results from the accelerated aging test and the long term test under in-use conditions ought to be established by mathematical models. It is also essential to, if possible, describe and give recommendations regarding all circumstances that might effect the durability of the studied product such as the maintanance, the manufacturing process, the design etc.

6 GENERAL RESEARCH NEEDS

The vital research needs of service life prediction have been described in general terms as barriers in section 3 of this paper. The need of a systematic approach for treating the evaluation of building materials durability has been examined by the review of some examples. The dignity of many research needs is best understood when dealt with at the level of a specific material or family of materials, but it is worthwhile to examine some of the general requirements in further detail.

6.1 The Gathering of In-Service Performance Data

The importance of reliable data on the in-service performance of materials is evident. Such data can be obtained basically in two ways: through exposure of test samples and, if the material has been used in buildings for a period of time, through field inspections of buildings. Standard procedures for the exposure of test samples at outdoor weathering sites are available, but there is a great need for procedures for feedback of the performance of materials under in-use conditions.

Building failure investigations can give information on the in-service performance of materials and components, but one has to bear in mind that most failures are caused by incorrect design or construction work or by faulty maintenance. Failure studies generate important data on the quality of design, construction and management of buildings but do not necessarily give the required feedback of information on the materials performance under normal conditions. But, systematic collection of failure data is important, as a complete prediction of service life ought to include an assessment of the failure risk due to design or specification errors and site errors. Procedures and methodologies for the collection of failure data have been developed and are being used by different organizations and institutes, but there is still a need for increased activities.

Field inspections of buildings as a means to generate service life data is an area in which research needs are apparent. Methods for inspecting the state of existing buildings or structures to assess the remaining service life of materials should be developed. This also includes the need of developing techniques that can be used in field inspections to evaluate degradation indicators.

Field surveys of random samples of buildings allowing statistical generalization of the state of the buildings and of their structural parts is a methodology that might be further developed. The method is suitable for descriptive investigations where the purpose is to make a statement on the situation for a defined population of buildings. The technique has been used by Sjöström et al (10) for studies of building failures.

6.2 The Description of the Environmental Affecting Building Materials

Poor knowledge of the environmental actions affecting building materials under in-use conditions is, in many cases, one of the most severe shortcomings in service life studies. Regarding measurement and reporting of climatological data and the classification of climates, several international activities of relevance are going on (3). The need for knowledge of the micro-environment and of the interdependence of micro-, meso- and macroclimates is great and work in this area should be accelerated. Data are needed on the environment within structures, such as information on moisture content and moisture transport mechanisms.

6.3 The Modeling of Degradation

Mathematical modeling of the observed degradation is an essential part of a service life prediction and, as a research area, this task is of growing interest. The models can be either deterministic or probabilistic. In an overview of models relating materials degradation response to the imposed stresses, Martin (11) chooses to present techniques based on reliability theory because of the often observed large variance in properties of building materials. This is a convincing argument and it might be added that the variations caused by the materials treatment during the erection and use of a structure underlines the need of a probabilistic view on the problem. Such an approach should also affect the work on creating the data to be used in the models, i.e. the variability of the different important parameters ought to be described.

7 CONCLUSIONS

The many examples of systematic methodologies for addressing research on service life show the need of such frameworks. A uniform methodology with an international acceptance would make research cooperation easier. Some of the reviewed methodologies specifically point out the need of documenting the observed degradation by mathematical models. Probability theory is, in this case, of special interest. It may also be beneficial to adopt a probabilistic view on service life studies as a whole by striving for establishing knowledge on the inherent variability of the parameters affecting the service life of building materials.

REFERENCES

1. CIB W80/RILEM 71-PSL. Prediction of service life of building materials and components, draft report, L. W. Masters, ed., (National Bureau of Standards, 1984).
2. Sneck, T., RILEM and durability. Matériaux et Constructions, Vol. 14, No. 83, (1981), 379 - 390.
3. Sneck, T., ed. Activities on durability of building products. Background study. (Espoo, Technical Research Centre of Finland, 1982).
4. Masters, L. W., Service Life Prediction: the Barriers and Opportunities, Proceedings of the 9th CIB Congress (Gävle, National Swedish Institute for Building Research, 1983).
5. ASTM E632-81, Standard Practice for Developing Accelerated Tests to Aid Prediction of Service Life of Building Components and Materials. (Philadelphia, American Society for Testing and Materials, 1982).
6. Coulbert, C. D., The Application of Encapsulation Material Stability Data to Photovoltaic Module Life Assessment. (Pasadena, Jet Propulsion Laboratory, 1983).
7. Australian Standard 1745, Part 2 - 1975, Outdoor Weathering of Plastics in the Australian Environment, Part 2, Guidance for Design Purposes. (Sydney, Standards Association of Australia, 1975).
8. Australian Standard CK 24, Part 1 - 1972, Outdoor Weathering of Plastics in the Australian Environment, Part 1, Commercial Products. (Sydney, Standards Association of Australia, 1972).
9. Blach, K., Brandt, E., Making the concept of durability manageable, Contribution to a CIB W60 symposium. (Hoersholm, Danish Building Research Institute, 1982).
10. Sjöström, Ch., Svennerstedt, B., Tolstoy, N., Defects of building materials and components in Swedish dwellings built 1955-79. Proceedings of the 9th CIB Congress (Gävle, National Swedish Institute for Building Research, 1983).
11. Martin, J. W., Time transformation functions commonly used in life testing analysis. Durability of Building Materials, Vol. 1, No. 2, (1982), 175 - 194.

BARRIERS TO THE PREDICTION OF SERVICE LIFE OF POLYMERIC MATERIALS

Ph. Eurin (1), J. Ch. Maréchal (2), R. Copé (2)

(1) Scientific Adviser, Building Research Projects, Ministère Industrie et Recherche, Paris
(2) CSTB, Division of the Physics of Materials, Grenoble, France

ABSTRACT

The principal barriers to the prediction of service life of polymeric materials used for buildings can be classified according to three categories:

- General barriers intrinsically associated with the problem of natural weathering;
- Barriers associated with polymeric materials themselves; and
- Barriers associated with the methodology selected for evaluation of the service life.

This paper gives a short description of these barriers through some examples, according to a general framework proposed for service life prediction of a material or a product.

1 INTRODUCTION

Over several centuries, the construction of buildings made use of a traditional know-how, based on the use of materials which resist the test of time and on the acquisition of exploitation techniques for these materials. Thus our ancestors, owing to their work with stone, wood, slate, fired clay, etc... achieved remarkable constructions, which serve as references for the present day, both architecturally and as far as quality of production is concerned.

At the present time, the large scale construction of housing at moderate cost has become a necessity. Construction techniques have therefore undergone considerable evolution, and the building trade increasingly calls upon new materials and makes use of industrially produced components.

However, this evolution towards industrialization and the utilization of new materials should not compromise the quality and the durability of buildings, which in Europe retain a considerable patrimonial value.

For this reason, it appears increasingly necessary for methods of evaluating the quality and service efficiency of building materials to be available. This applies particularly to new materials, such as the plastics, which have recently appeared and which, for the most part, do not have sufficiently long standing references for extensive utilization.

Predicting the service life must be the result of a systematic and scientific approach, but the application to real cases comes up against various barriers which this contribution aims to consider. In 1981, the CIB and the RILEM established a working committee, commissioned to establish a report on the methods currently used to study the durability of materials and constituents for buildings, and to propose a systematic methodology for predicting their service life.

The different investigational areas are detailed in the following synoptic table (fig. 1). These include in particular:

1) The determination of functional requirements and their limits

2) Identifying the environmental constraints

3) Characterization of the material (composition, structure, performance, characteristics)

4) Understanding the mechanisms of degradation

5) Selection of a degradation index

6) Definition of an artificial aging test

7) Correlation between natural weathering and artificial aging and

8) Extrapolation in time prediction of service life.

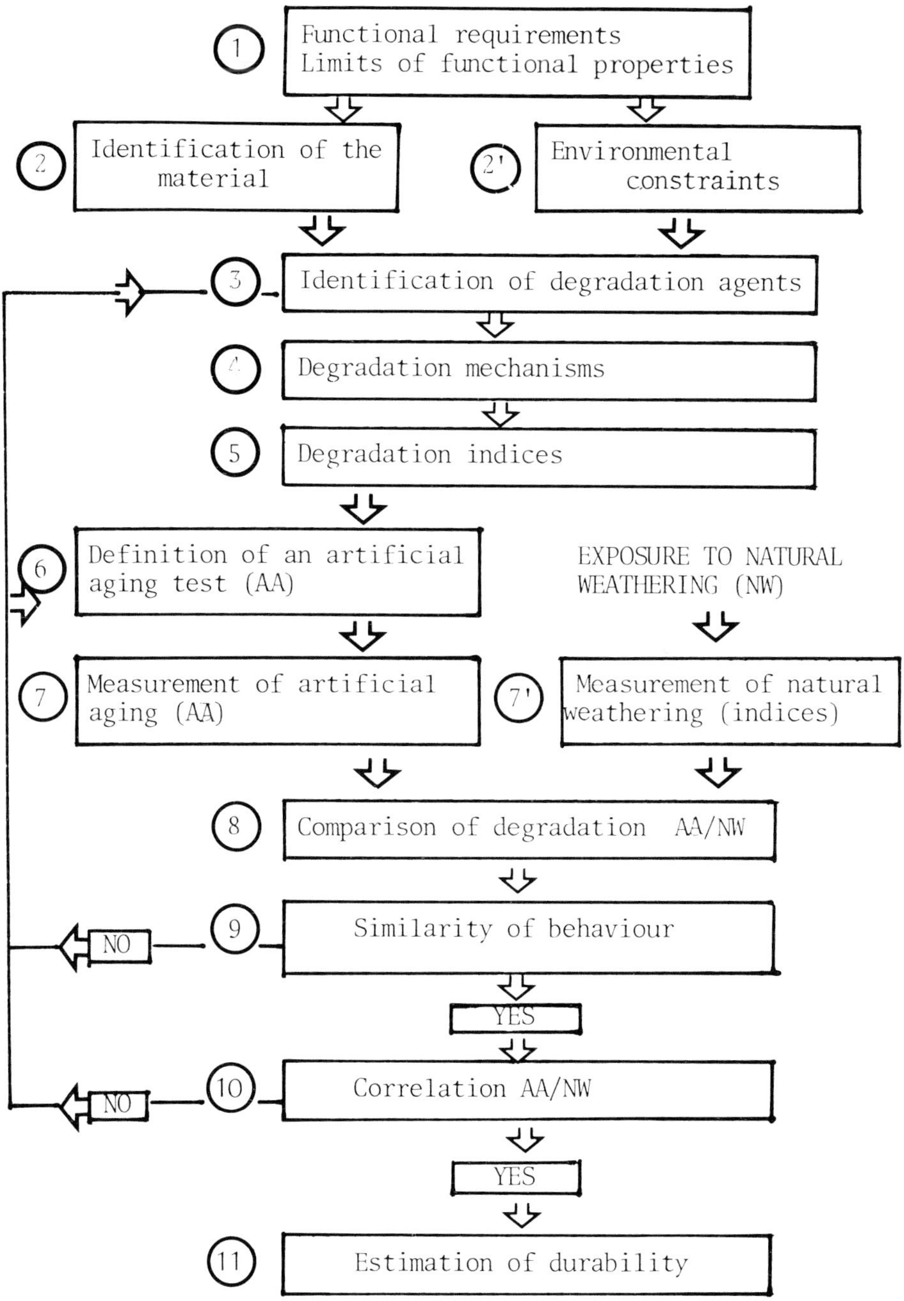

Figure 1 - Framework for service life prediction of a material (a product) -

2 CHOICE OF FUNCTIONAL CHARACTERISTICS AND THEIR LIMITS

For a given purpose, a material must simultaneously satisfy several functions owing to its own characteristics. The mechanical behaviour, thermal properties, appearance, as well as the general physicochemical properties, constitute a large range of parameters which are often difficult to observe in the laboratory, taking into account the theoretical as well as experimental means likely to be available.

Moreover, the service life of a material, that is its capacity to meet the requirements for a given purpose in normal operating conditions, may be limited to maintaining several functional characteristics at an adequate level.

The choice of functional characteristics is based on an approach that will always retain a somewhat arbitrary dimension.

This choice thus constitutes an important first barrier, in consideration of the fact that assessing the service life of a material can vary appreciably according to whether reference is made to one or other of these characteristics. The behaviour of PVC boarding over time provides an example of this first problem. A slight variation in appearance (color), which would indeed only constitute a second order defect if uniform, can become unacceptable if not uniform, while the mechanical characteristics retain a satisfactory level of performance. The service life regarding the "appearance" feature is thus very different from the "mechanical" service life.

The experimental course of a functional characteristic over time can be illustrated graphically by a curve F(t) (fig. 2). The intersection of the curve with the straight line parallel to the x-axis, describing the threshold F_S beyond which performance is inadequate, enables the definition of the point on the abscissa T. This may be considered as the service life of the material.

Taking into account the general profile of certain curves $F_1(t)$, very often asymptotic when $t \to \infty$, a small variation ΔF_S is likely to generate a very great Δt; for this reason, the fairly precise definition of the threshold F_S constitutes a delicate choice, all the more delicate when the functional characteristic F_S is dependent on an approach that is not completely objective (appearance, color...).

Other functional characteristics whose observation is essential may present evolutions over time with profiles of the type $F_2(t)$ (fig. 2); this is illustrated by the variation of the tensile impact test resilience for rigid PVC with respect to time (1).

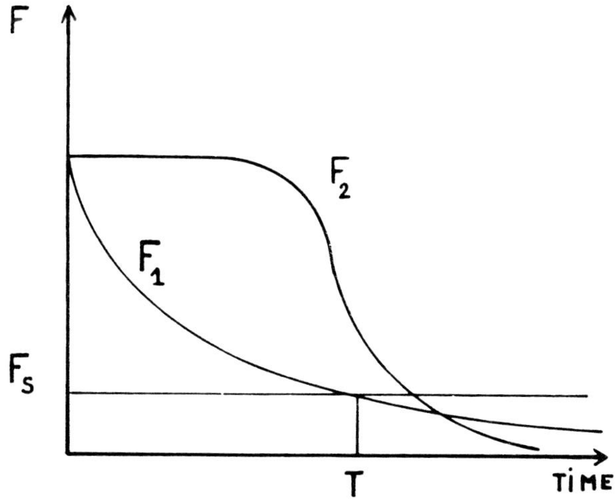

Figure 2 - Schematic evolutions of functional properties versus time.

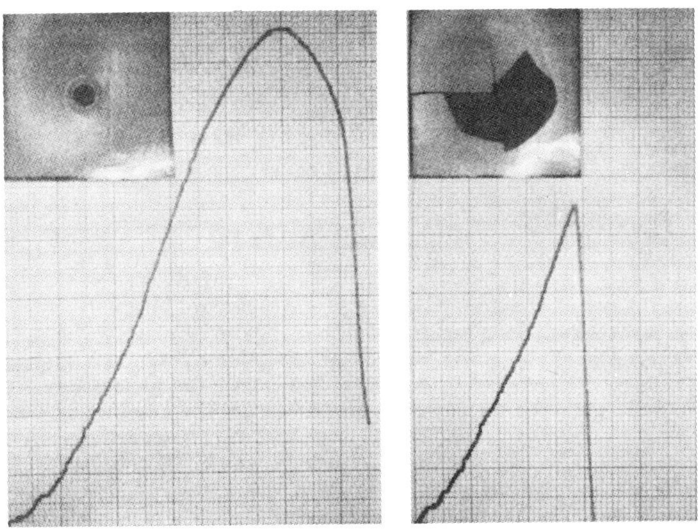

Figure 3 - Strength vs. displacement curves for ductile (a) and brittle (b) samples.

Such curves render virtually impossible any long term extrapolation which would enable definition of the service life of a material or product on examination of that particular property.

3 UNDERSTANDING THE ENVIRONMENTAL CONSTRAINTS ACTING ON THE MATERIAL IN USE

It is difficult, if not impossible, to establish an exhaustive list of the factors at the origin of the environmental constraints to which the material in service is subjected, owing to the great variability of the environment in nature, in intensity, and also sometimes in predictability (conditions with a low occurrence probability can generate large scale damage).

A list of these degradation factors, for a "building" environment, is proposed in table 1, which is taken from ISO DP 6241. We will not present a detailed description of each factor, but will attempt to underline the interactions, the synergies or simply the peculiarities of certain situations.

The permanent, periodic or often unpredictable mechanical stresses are particularly difficult to analyse in the case of polymeric materials. This is particularly true for polymers whose rheological behaviour is of the viscoelastic type and is thus simultaneously a function of stresses (σ), deformations (ε), temperature and time through the derived fuctions $\frac{d\sigma}{\partial t}$ or $\frac{d\varepsilon}{dt}$. The embrittlement of PVC at low temperature or at a high rate of strain constitutes a good example for this point (fig. 3). The same is true as far as fatigue behaviour is concerned, which for polymers, in particular, must be considered in terms of both dynamic and static fatigue.

It is also important to underline that heat factors can relax or generate their own inherent mechanical stresses. The relaxation of internal stresses, produced by the mode of transformation of the product or by its manner of application in the building, can result in a deformation (expansion, contraction, warping) which will itself generate new stresses as, for example, at the level of attachment points.

The action of mechanical factors can be appreciably intensified by chemical factors or by radiation:

- for polyolefins, water can favour the development of cracking under mechanical strains of small amplitude,
- the action of oxidizing agents (oxygen, ozone) on the degradation of elastomers under stress is also well known (fig. 4),
- mineral deposits on the surface of certain asphaltic sealing coatings are likely to constitute an embrittlement

NATURE	ORIGIN			
	EXTERNAL TO THE BUILDING		INTERNAL TO THE BUILDING	
	Atmosphere	Ground	Occupancy	Design consequences
1. Mechanical agents				
1.1 Gravitation	snow loads, rain water loads	ground pressure, water pressure	live loads	dead loads
1.2 Forces and imposed or restrained deformations (4.01)	ice formation pressure, thermal and moisture expansion	subsidence, slip	handling forces indentation	shrinkage, creep, forces and imposed deformations
1.3 Kinetic energy	wind, hail, external impacts	-	internal impacts, wear	water hammer
1.4 Vibrations and noises	wind, thunder, airplanes, explosions, traffic and machinery noises	earthquakes, traffic and machinery vibrations	noise and vibration from music, dancers, domestic appliances	servies noises and vibrations
2. Electromagnetic agents				
2.1 Radiation	solar radiation, radioactive radiation	-	lamps, radioactive radiation	radiating surface
2.2 Electricity	lighting	stray currents	-	static electricity, electrical supply
2.3 Magnetism	-	-	magnetic fields	magnetic fields
3. Thermal agents	heat, frost, thermal shock	ground heat, frost	user emitted heat, cigarette	heating, fire
4. Chemical agents				
4.1 Water and solvents	air humidity, condensation, precipitations	surface water, ground water	water sprays, condensation, detergents, alcohol	water supply, water waste, seepage
4.2 Oxidizing agents	oxygen, ozone, oxides of nitrogen	-	disinfectant, hydrogen peroxide	positive electrochemical potentials
4.3 Reducing agents	-	sulphides	agents of combustion, ammonia	agents of combustion, negative electrochemical potentials
4.4 Acide	carbonic acid, bird droppings, sulphuric acid	carbonic acid, humic acids	vinegar, citric acid, carbonic acid	sulphuric acid, carbonic acid
4.5 Bases	-	lime	sodium hydroxide potassium hydroxide, ammonium hydroxide	sodium hydroxide, cement
4.6 Salts	salty fog	nitrates, phosphates, chlorides, sulphates	sodium chloride	calcium chloride, sulphates, plaster
4.7 Chemically neutral	dust, soot	limestone, silica	fat, oil, ink, dust	fat, oil, dust, soot
5. Biological agents				
5.1 Vegetable and microbial	bacteria, seeds	bacteria, molds, fungi, roots	bacteria, house plants	-
5.2 Animal	insects, birds	rodents, worms	domestic animals	-

TABLE 1 - Degradation factors (agents) relevant to building performance -

layer for the product; cracks and microcracks initiated in this superficial layer can propagate throughout the entire thickness of the product.

Research investigating natural weathering of organic materials outdoors are also well acquainted with the considerable acceleration in degradation which results from different climatic agents and from solar radiation, particularly when samples are exposed under mechanical stresses.

The climatic stresses can be considerably influenced in the same geographical location - that is integrating the existence of a microclimate or other parameters such as the altitude or the site - by the position and the orientation of the constituent. Stagnation of water, organic or mineral deposits, and the development of micro-organisms are indeed more likely on a horizontal than on a vertical surface. In the same way, a North facing or a South facing plane will be subjected to very different solar radiation energy and hygrothermal conditions. Figure 5 details the solar energy reaching a vertical plane according to its orientation on an open site at latitude 46°N.

The heat environment of polymeric siding installed over an air space or over an insulator, as well as the temperature profile of a sealing coating on an inert structure or on a heat insulator, constitute other examples of effect of materials and design upon the service conditions (fig. 6).

4 IDENTIFICATION OF THE MATERIAL

Identification is an important stage in the organization of durability studies. Degradation phenomena can only be evaluated by reliable scientific approaches if the composition or structure of the material actually employed is known. Identification of a material is also fundamental for certification purposes. Even if the scientific and technical information for this may be more concise, it is essential to have a true "finger-print" available, providing confirmation that the material actually employed corresponds to the reference material which has been previously investigated in an evaluation, approval or conformity procedure.

Within this approach to identification of a material, four barriers can be underlined as important:

1 - The chemical composition of a material is generally an industrial secret: for this reason, it is frequently not disclosed by the manufacturer. Laboratories must therefore set up sophisticated experimental procedures, which are generally costly and complex, in order to have at their disposal details of the formulations. These theoretical, as well as

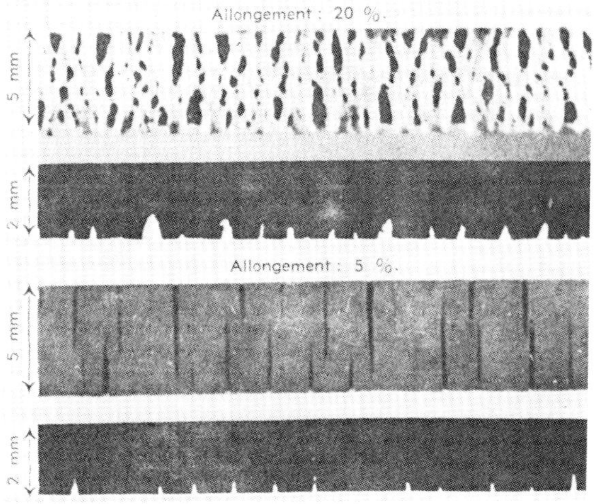

Figure 4 - Measurement of the ozone degradation considering the edges of the elastomeric samples.

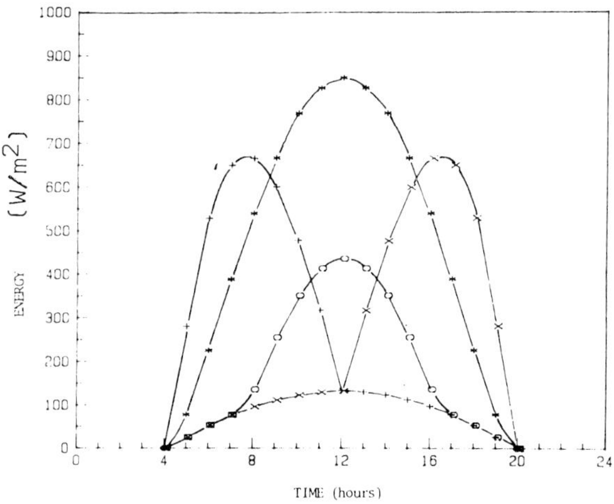

Figure 5 - Instantaneous solar radiation energy reaching some surface versus time and direction -
Date : June 21 - Latitude : 46°N - ordinary conditions of sky

* horizontal surface - x western vertical surface - o southern vertical surface
x eastern vertical surface

experimental tools for analysis often exist only in basic research institutes.

2 - If the nature of the basic polymer constituting the material is often well known, the additives are less so. Nevertheless, the influence of these additives is decisive for the quality and durability of the material. These additives, which can be formulation agents (thermal stabilizing agents, UV, impact etc...) but also transformation agents (processing aid, thermal stabilizing agents...) are introduced in very small quantities into the material. A quantitative, as well as qualitative analysis, would be required for characterization. How fine and precise must this analysis be? The reply is of course specific to the polymer family investigated.

3 - During its transformation to product form, the matter is subjected to a certain number of sometimes poorly understood thermo-mechanical stresses, which by acting on the structure or on the composition ("consumption" of stabilizing agents), can appreciably modify the durability of the product. Investigations (3) have, for example, shown that PVC sections of the same formulation, extruded on the same machine, according to an appreciably different thermal extrusion profile did not have the same behaviour in service; the appearance, and in particular the colour, showed degradation (pinkening phenomenon) after being in service outdoors as boarding for less than two years. In addition to manufacturing conditions, which are sometimes poorly controlled, are problems likely to arise from particular handling of products during storage on trading sites or on work sites before putting them into service (for example, storage exposed to the elements without protection, sometimes for long periods, of products for indoor use) or during transportation or handling concerned with transportation (impact, static fatigue, etc...).

4 - The difficulties that certain laboratories encounter in identifying a material also pose, at a secondary level, the problem of quality control, which must necessarily rely on an experimental procedure that is easy to put into practice, but is nevertheless reliable. The definition of such an experimental procedure represents a real barrier to be overcome by the institutes on the one hand, but also by the transformers (small industrial companies) for their own monitoring system. For this reason, the "finger print" established using certain well-selected specific characteristics, while still respecting the industrial secret of the formulation, can be an especially useful identification element within the framework of quality control procedures.

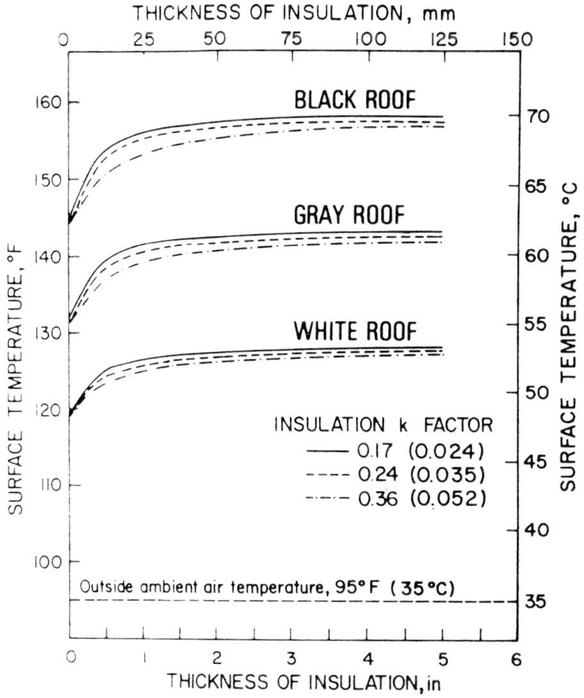

Figure 6 - Surface temperature of the roof membrane as a function of the amount of insulation for the parameters given in table 3 (ref. 2).

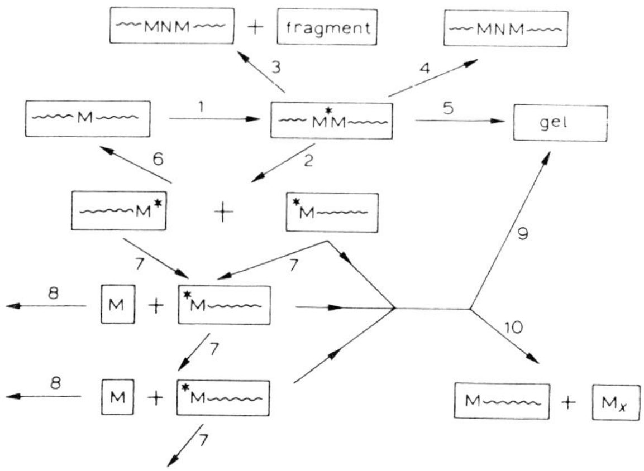

Figure 7 -

General polymer degradation scheme. 1, Production of active site by heat, irradiation or mechanical means; 2, chain scission; 3, loss of side group; 4, transformation of side group; 5, cross linking; 6, cage recombination; 7, depropagation; 8, diffusion of monomer; 9, intermolecular abstraction leading to cross linking; 10, intramolecular abstraction (back biting) leading to oligomeric products. Note that reactions with impurities or additives have been omitted.

5 UNDERSTANDING THE DEGRADATION MECHANISMS

Understanding the degradation mechanisms is of first importance, as they govern the evolution of the material and its service life.

These processes are generally extremely complex. They are the resultant of a series of sometimes unknown elementary mechanisms, which are themselves dependent upon numerous factors:

- the nature of the basic polymer,
- additives (antioxidants, stabilizing agents, fillers),
- manufacturing impurities,
- transformation defects,
- environmental parameters.

These are considered to be the principal agents of polymer degradation (fig. 7). (4)

The fundamental mechanisms relative to a degradation factor can be further considerably perturbed by the simultaneous action of another factor, the synergetic effects being particularly different to analyse at the reaction mechanism level. Once the mechanisms have been identified, other difficulties remain, in particular the analysis of the relations of causality between the basic mechanisms and the variation of the functional properties.

Oxygen, when combined with heat or with solar radiation, generates reactions of thermo-oxidation or photo-oxidation whose damaging effects on polymers are well-known and include embrittlement, colour change, degradation of appearance and loss of initial mechanical performance. Each polymer reacts differently to these constraints.

<u>Polyolefins</u> are sensitive to oxidation; the mechanical properties depend to a great extent on the oxidation of the resin. Research has shown that the choice of carbonyl index can be enough to observe the state of polymer degradation.

The polymeric radicals, formed either thermally or photochemically, react with oxygen to give hydroperoxides.

$$R^{\bullet} + O_2 \rightarrow ROO^{\bullet}$$
$$ROO^{\bullet} + RH \rightarrow ROOH + R^{\bullet}$$

Each molecule of hydroperoxide undergoes photolysis which results in the creation of ketone.

$$ROOH \rightarrow RO^{\bullet} + {}^{\bullet}OH$$
$$RO^{\bullet} + RH \quad ROH + R^{\bullet}$$

RO: → R' - C = O
R''

Then the Norrish reactions contribute to the creation of new radicals associated with the breaks in the chain.

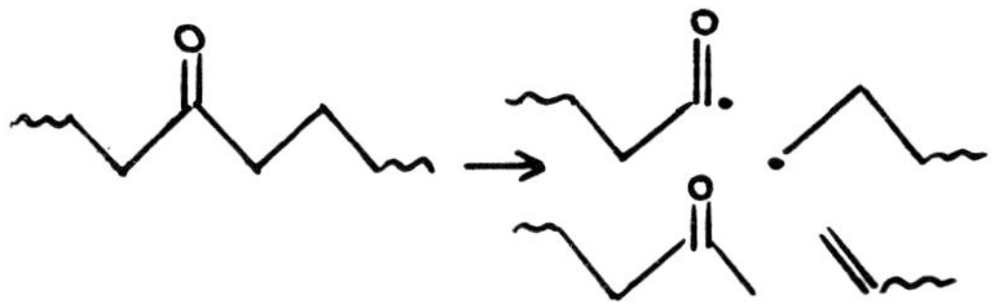

Norrish I

Norrish II

The loss of mechanical properties is the macroscopic consequence of all these elementary reactions which ultimately explain the poor long term behaviour of low stabilized polyolefins.

Degradation of polyvinylchloride (PVC) results from a different process. It has been established that the chemical phenomenon tracing the degradation is the loss of hydrochloric acid in the polymer chains, whether the cause be thermal or photo-chemical.

The dehydrochlorination mechanism begins by an initiation reaction at the weak points in the structure of the polymers (branching, non-saturation, oxidized groups) or at impurities.

$$- CHCl - CH_2 - CHCl - CH_2 - CHCl - CH_2$$

$$\downarrow HCl$$

$$- CH = CH - CHCl - CH_2 - CHCl - CH_2$$

$$\downarrow HCl$$

$$- CH = CH - CH = CH - CHCl - CH_2$$

This mechanism explains the formation of hydrochloric acid and the colouring of weathered PVC due to the polyenyl structures.

The oxidation reactions and the intermolecular reactions will then break the polymer chains, and will lead to a decrease in the mechanical performance.

$$-(CH{=}CH)_x \xrightarrow{O_2} (CH{=}CH)_y - CO - CHCl - CH_2 -$$

The degradation reactions of PVC are now better understood, but still remain highly complex and controversial. They have nevertheless enabled a weathering index to be proposed, based on the determination of the induction time for deshydrochlorination to follow the weathering damages.

Glass-fibre reinforced polyesters (GRP) (5) behave differently from the two previous polymers, as the weathering is manifested by:

- superficial degradation, resulting from a chemical effect (hydrolysis of the resin) and a mechanical effect (erosion) of water,

- destruction of bond between the fibers and the resin.

Once diffused in the resin, the water may act as a plastifying agent or may cause the polymer to swell, and attack the ester bonds thus reducing the reticulation rate.

Passage of the water to the heart can also be facilitated by hydrolysis of the silane coupling agents between, the resin and the fibers. After damage to the bond between the two constituents, the fibers then behave as veritable wicks for the water. For example, the siloxane bond between resin and glass-fiber surface (through an oxide) can be attacked according to the reaction:

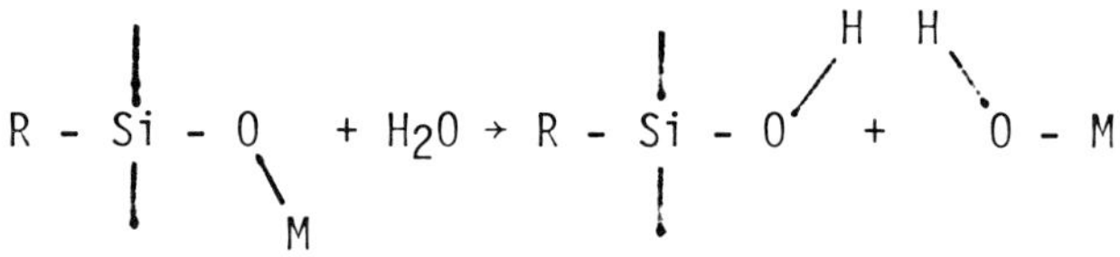

Elastomers (6) such as polychloroprene and EPDM are the most common in the manufacture of joints. They weather essentially by thermo-oxidation. The black carbon protects them from sunlight. The oxidation processes set off peroxide formation whose decomposition leads to a split in the chains, which in the early phases overrides reticulation. The decomposition of the alkoxy radicals gives rise to various oxygenated compounds (carbonyl, hydroperoxide). This classic degradation pattern results in reticulation of the elastomer, which thus loses its initial elastic properties.

The examples presented show clearly the value of a sound understanding of the mechanisms. They enable selection of accelerated aging tests, and the degradation index used for predicting the durability; they also help justify the correlation between artificial aging and natural weathering.

6 THE DEGRADATION INDICES

The study of weathering of materials is also dependent on a phenomenological approach for determining the degradation indices.

The term degradation index is applied to a measurable property, evolving progressively and continuously in time, according to simple laws which enable the most reliable extrapolation possible.

That property must also correlate with the essential functional characteristics of the materials.

Finally, on examination of these representative indices for each material, it must also be possible to propose correlations between the different modes of damage investigated (natural weathering and artificial aging). The choice of these specific indices for each material is in itself a significant barrier to be overcome, and examples will be developed in the following chapter.

7 ARTIFICIAL AGING TESTS

This is an essential part of investigations into the prediction of service life.

These tests are necessary because, in the development of new materials or new production techniques, it is not possible to wait for the results of natural weathering.

Above all, the first point to be emphasized is that it is technically impossible to artificially reproduce the climatic environment to which materials are actually exposed in service. Accelerated degradation tests are selected according to the reaction of each material with climatic factors. In particular, an attempt is made to reproduce artificially the action of the most aggressive degradation factors for the material investigated. Understanding the mechanisms is equally important at this stage, for if acceleration can be achieved by acting on the intensity or the amplitude of the factor, or even on its duration and its mode of application, these mechanisms should not be greatly modified as the acceleration of the degradation phenomena must result from increased reaction kinetics.

A second barrier lies then in the satisfaction of the preceding conditions.

For PVC, the main factor in weathering is UV radiation from the solar spectrum. The artificial aging mode selected to accelerate degradation of the material is, therefore, permanent exposure to UV radiation emitted by a filtered xenon lamp. The air atmosphere for the test is controlled for temperature and humidity. The degradation index, based on the chemical phenomenon of hydrochloric acid loss from the macromolecular chains (which is the tracer for weathering) is the deshydrochlorination index (DHC index) (fig. 8). It is the time required to detect HCl loss in well-defined

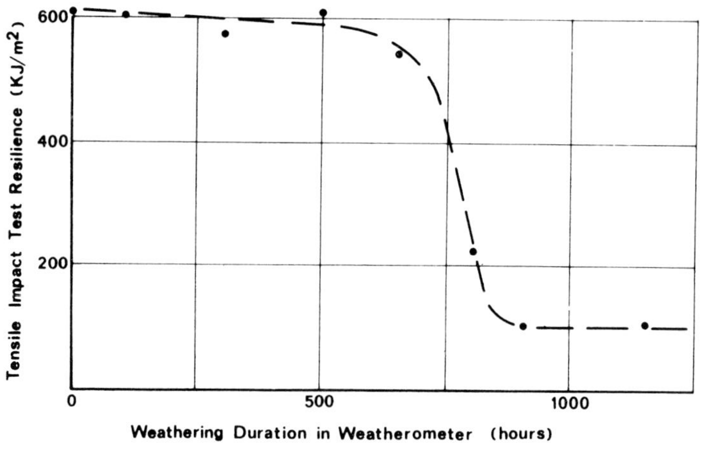

Figure 8 - Tensile impact test resilience of rigid PVC vs. weathering time in weatherometer.

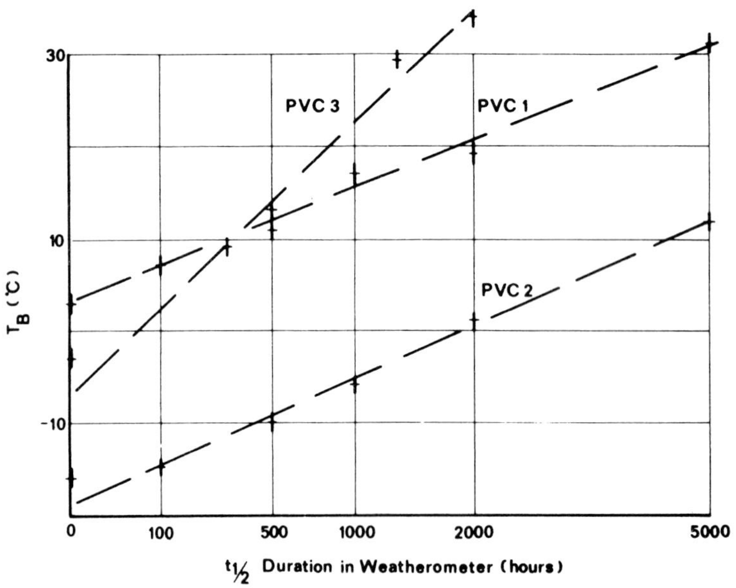

Figure 9 - Brittle temperature T_B vs. weatherometer time $t^{1/2}$ for three PVC samples.

test conditions. This index is associated with the quantity of residual thermal stabilizing agents in the material. (7)

Another index, the embrittlement temperature, may equally serve in predicting the durability of PVC (fig. 9) and is determined by impact perforation tests. The embrittlement can be explained by the creation of defects on the PVC surface, which become more important as weathering progresses.

For GRPs, the method of accelerated aging selected is boiling water, or a high temperature atmosphere (70 to 80°C) with high relative humidity (95 to 100% RH). For this material, water appears to be the most corrosive of the natural weathering factors by virtue of both its mechanical and chemical action.

It can be observed that the boiling water test acts in a similar way to natural weathering. The index selected is the rigidty modulus which decreases progressively over the degradation time. The chemical behaviour of the resin under torsion is not affected by the test. A qualitative correlation of the weathering profiles, for various formulations of GRP, has been obtained between the boiling water test and natural weathering.

For elastomers covered with carbon black, the heat degradation in a drying oven has allowed investigation of the range of joints used in building, with polychloroprene and EPDM base. The weathering indices selected (measurement of swelling and tensile relaxation) are used to follow the course of heat damage, and have served in correlation studies with natural weathering.

Such a heat test can be valid criterion in classifying joints of different qualities for resistance to thermo-oxidation.

8 CORRELATION BETWEEN NATURAL WEATHERING AND ARTIFICIAL AGING

This is without doubt the greatest barrier encountered in the study of the durability of a material. It has already been emphasized that it is impossible to achieve complete simulation by accelerating natural weathering; there is, therefore, always to a greater or lesser degree a modification of the degradation mechanisms artifically induced. Any correlation which, by virtue of the initial theoretical hypotheses (similar degradation mechanisms), should result from the simple comparison of reactional kinetics, has in fact proved to be extremely difficult. In certain cases, the phenomena may be sufficiently subtle for the choice of degradation index to be reconsidered.

PVC is a good example of such difficulties. For this reason, two indices have been proposed (deshydrochlorination index and ductile-brittle transition temperature of the material).

Another estimation of the correlation can result from an analytical approach and a mathematical model of the degradation phenomena.

However, in this case, certain problems arising from points discussed earlier should not be neglected (functional requirements, understanding of the environment, of the material, of the phenomena) and also certain others such as:

- the number of parameters to be treated simultaneously,
- effects of synergy,

etc ...

The complexity of this problem understandably leads to the development of mathematical methods which, with a stochastic approach, allow a theoretical behaviour to be understood by general models and methods of prediction to be improved (8).

9 CONCLUSIONS

It would appear then that the principal barriers to the prediction of durability in polymeric materials used for building can be classified according to three categories:

1 - General barriers intrinsically associated with the problem of natural weathering. The time scale will never be reduced. This is a sizeable drawback as the time factor intervenes at every level of evaluation of materials, just as it does in the reaction kinetics and the visco-elastic mechanical behaviour. The climate is essentially variable and is hardly predictable in considering the future.

2 - Barriers associated with polymeric materials themselves. Organic materials, by their very nature, exhibit a specific behaviour owing to their viscoelastic mechanical behaviour. This condition evolves with the time-temperature parameters. These two factors can provoke fatigue, flow or relaxation which is difficult to control, and is, thus, difficult to predict.

- Identification of materials.

Existing test methods and results of laboratory research must constantly be adapted to the problem of identification. Complex and costly operations using sophisticated equipment are sometimes required.

The úse of simple control tests associated with the existence of data banks concerning materials would enormously simplify the identification procedures.

• The degradation mechanisms.

This is one of the difficult points, whose importance should diminish with the progress in research on materials degradation.

3 - Barriers associated with the methodology selected for evaluation of the durability.

• The definition of functional requirements and of the limits of use is determined by convention.

It would be interesting to submit to the international associations the question of determination of a lower limit for the properties of polymeric materials used in buildings.

• The design of accelerated damaging test.

Taking into account that it is at present impossible to reproduce for each material all the components of natural weathering with the same degree of acceleration, simplified accelerated tests are required to develop degradation models and to reveal the different qualities and improvement of formulations.

Unfortunately, acceleration introduces inevitable distortions in the degradation processes. This makes monitoring of accelerated results against natural weathering essential.
The solution of problems concerning durability, which requires long-term testing and a scientific approach to the degradation phenomena, lies in the research work carried out by the institutes investigating building materials. As L. W. MASTERS proposes, greater cooperation between research centers is required to solve these questions and to overcome these barriers (9).

REFERENCES

1. G. ROUX, Ph. EURIN, J. Macromol. Sci. Phys. B20(4), p 505 (1981)
2. W.J. ROSSITER, R.G. MATHEY, NBS IR 76-987, Feb. 1976
3. Contrat DGRST 74-7-0162
4. H.H.G. JELLINEK, Aspect of degradation and stabilization of polymers, Elsevier S.P.C.
5. R. COPE, Cahiers du CSTB n° 1426, Mars 1977
6. J.L. CHEVALIER and al, Cahiers du CSTB n° 1616, Déc. 1979
7. J.Ch. MARECHAL, Cahiers du CSTB n° 1589, Sept. 1979
8. J. MARTIN, in 2nd Int. Conf. on the Durability, of Building Materials and Components, Sept. 1981
9. L.W. MASTERS, the 9th CIB Congress Stockholm, Aug. 1983

BARRIERS TO THE PREDICTION OF SERVICE LIFE OF INORGANIC MATERIALS

Micheline Regourd

Lafarge Coppee Recherche
Trappes, France

ABSTRACT

The service life prediction of inorganic materials is, at present, primarily deduced from studies on mechanical and chemical performance in given environments. Selected models of the behaviour of concrete structures are presented. They are either structural engineering models or materials science models. They can predict mechanical properties, frost resistance or behaviour regarding carbonation. The barriers to service life prediction are related to the complexity of multicomponent systems, to little known degradation mechanisms, to unsatisfactory accelerated tests and to insufficient preventive measures. Examples of chemical deterioration of concrete are given. They emphasize the need for more research in materials science and a closer collaboration between structural engineers and materials scientists.

1 INTRODUCTION

The service life prediction of inorganic building materials i.e. natural stones, bricks and concretes is usually deduced from:

- observations of old structures
- studies of deterioration mechanisms on laboratory specimens or core samples extracted from damaged structures
- modelling of materials or structures behaviour in special environments.

These predictions, based mainly upon chemical durability and mechanical performance, have resulted in specifications listed in many standards and codes of practice. However, recent cases of unexpected degradation have ratified:

- the deficiency of our knowledge of the chemical mechanisms of attack on materials
- the inadequacy of simple models, which take into account too few degradation parameters
- certain unsatisfactory accelerated tests and
- a lack of close collaboration between structural engineers and materials scientists.

Inorganic non-metallic materials (Table 1) include, among other materials, cementitious composite materials, such as mortars and concretes; these are considered as living products because their structure varies with time due to both the cement hydration and the cement paste-aggregate bonding. So the general service life prediction of concretes is not yet possible due to the great complexity of multi-component systems comprising concretes. This paper will address some of these complex systems.

Table 1. Inorganic non-metallic building materials

Natural stones

Manufactured products

- ceramics: terra cotta, earthenware
- glasses
- bricks, adobes

Cementitious materials

- portland cement clinkers
- slags: blast furnance slags
- pozzolanas: natural volcanic rocks, fly ash, silica fume, rice husk ash
- fillers: limestones, marls
- gypsum, phosphogypsum

Composite materials

- mortars
- concretes
- masonries
- plasters

2 MODELS OF CONCRETE BEHAVIOUR

The structure of concrete composed of cement, sand and aggregates has been considered at different levels (1, 2):

- at the macroscopic level. It is constituted of the cement paste matrix embedding sand grains and coarse aggregates. Models such as for time-dependent behaviour (3), are structural engineering models using finite elements. They are rather simple, looking upon concrete as a quasi-homogeneous material.

- at the microscopic (or meso) level. The matrix appears heterogeneous, composed of calcium hydrates (C-S-H, $Ca(OH)_2$, sulfoaluminates), anhydrous cement grains (clinker, slag, pfa), and capillary pores with various amounts of water. Sand grains and aggregates occur as polyphased minerals that are more or less porous. Models are materials engineering models built on pore distribution or polygonal inclusions. Numerical or stochastic, they have been developed mainly for predicting the mechanical performance of concrete.

- at the submicroscopic level. The cement hydrate, C-S-H, is colloidal-like. Models of its structure are composed of lamellar sheets in an irregular array (6,7). Water molecules are more or less absorbed at the surface sheets depending on the relative humidity. There are two kinds of pores: gel (small) pores and or capillary (larger) pores. These are materials science models.

2.1 Prediction of Mechanical Properties

Mechanical properties of concretes depend upon (i) the mechanical properties of the cement paste and aggregates and (ii) the cement paste to aggregate bond. Compressive and flexure strengths, elastic deformations, shrinkage, swelling and creep are all characteristics that are due to externally applied stress or to moisture transfer; all these have been tentatively modelized. While all authors agree on the influence of porosity, however, some use the volume fraction of total pores (8), some use volume fraction of the capillary pores (9) and others use the size of the largest pore (10). Fracture mechanics has been applied to the cracking process in concrete under loading (4). The role of the cement paste-aggregate bond has been clearly shown and includes the mineralogy and the roughness of aggregates (Figs. 1 and 2).

2.2 Prediction of Service Life Regarding Frost Action

In northern countries, frost action is one of the major causes of concrete degradation. Several mechanisms have been proposed (11, 12, 13), including hydraulic pressure, osmotic pressure or moisture transfer. Two characteristics can be measured:

the spacing factor $\overline{L}$ (14) and the critical degree of saturation Scr (15). The spacing factor, related to the distance between two air voids on a polished section of concrete before freezing, has been limited to less than 200 µm. This value was obtained in the case of severe treatments (14). In fact, recent studies have shown that:

- natural freezing and thawing cycles are less harmful than shown by the ASTM test and $\overline{L}$ can be as high as 600 µm (16)
- the use of water reducer increases the limit of $\overline{L}$ (17) and
- the determination of Scr, by measuring the residual dynamic E modulus, leads through a geometrical relation to values of $\overline{L}$ which are greater than 200 µm (18).

The potential service lifes calculated by Fagerlund (18) for concretes of different water/cement ratios are as follows:

TIME	W/C
1.9 years	0.573
41.0 "	0.50
835.0 "	0.45

The above service lifes were projected assuming each of the concretes had the following other characteristics:

- cement content: 298 kg/m3
- degree of cement hydration: 0.80
- total air content: 0.071
- critical degree of saturation: 0.80
- spacing factor equivalent: 400 µm.

These first results were published in 1981. Perhaps new evaluations will be presented at this seminar taking into account more variable parameters, such as new air entraining agents, new cement types, moisture content and nature of aggregates. Salt scaling is another parameter in the frost action on concrete that is not entirely estimated. It has already been related to salt concentration and moisture content.

2.3 Prediction of Service Life Regarding Carbonation

An analytical model has been described by Houst, Roelfstra and Wittmann (19). If the concrete composition is known, the service life of a structure exposed to an arbitrary climate can be predicted. The mutual interaction of drying, diffusion of carbon dioxide, the formation of carbonate, a function of the degree of carbonation and the CO_2 gas concentration have been expressed in equations. Results of a numerical evaluation of the set of equations are in good agreement with experimental data. After an

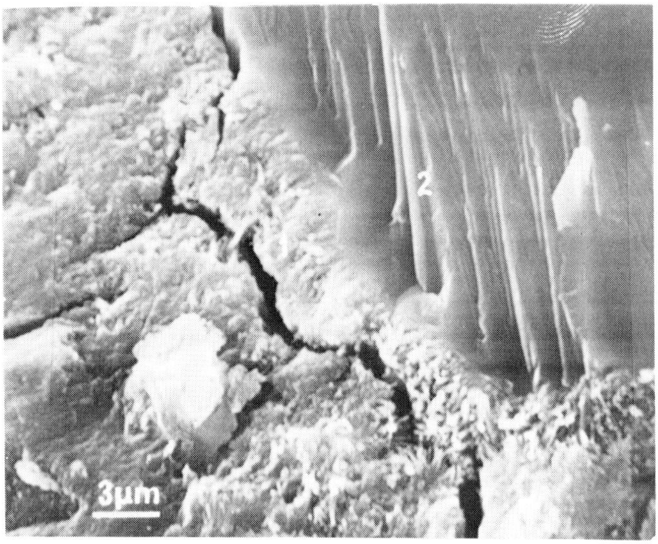

FIG 1. Cement paste - aggregate bond: crack in cement paste
1. C-S-H
2. Calcite aggregate

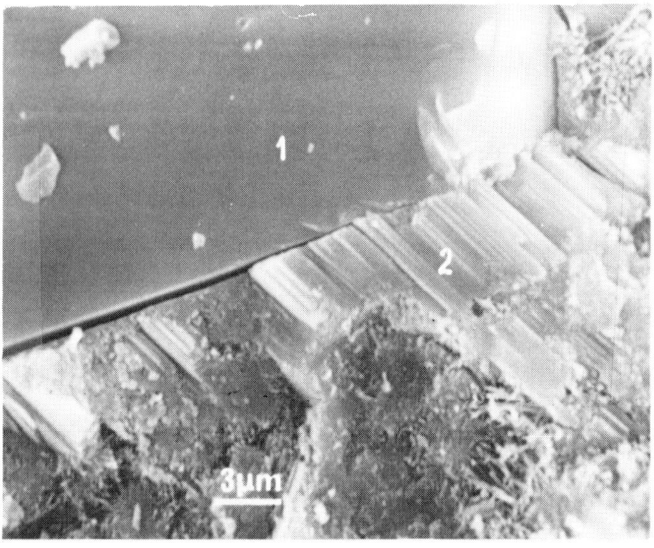

FIG 2. Cement paste - aggregate bond: crack at the interface
1. Siliceous aggregate
2. $Ca(OH)_2$

initial rapid carbonation, the growth of the carbonated layer follows approximately a (t) - 1/2 law.

After a given time t, a rough estimation of the depth of carbonation is given by the following equation

$$d = a + b\sqrt{t}.$$

This simple equation had already been suggested after an empirical study by Vénuat and Alexandre (20). The prediction of mechanical properties and of service life regarding either frost action or carbonation are first positive steps in the evaluation of long term concrete behaviour but barriers still exist to this prediction and, more particularly, in cases related to the physicochemical durability.

3 BARRIERS TO SERVICE LIFE PREDICTION

Barriers to the service life prediction of inorganic building materials can be related to:

- unknown or little known mechanisms of degradation,
- unsatisfactory accelerated tests and
- insufficient preventive measures.

Mineral building materials can be attacked by chemical, biological, physical and mechanical agents (Table 2).

Building materials are frequently porous. Chemical corrosion processes involve fluid transport, reactions at solid-fluid interfaces, capillary pressures and crystal growth. External signs of deterioration appear as surface erosion, crazing, pitting, peeling, cracking, flaking, discoloration, salt crystallization and gel extrusion.

Waters, soils, atmospheres and bacteria can be aggressive agents to concrete. Their action depends upon the concentration of harmful ions, humidity and temperature (Table 3) and on the quality of materials related to the porosity, permeability and surface area themselves in relation with mix proportions, cement quality type of aggregate and strength characteristics.

The more common corrosive ions to building materials are chloride, sulfate and carbonate. New results have recently appeared on the chloride diffusion (22) and ettringite expansion, both corresponding to external attacks and on the dissolution of porous silica in relation with the alkali-aggregate reaction in concrete, and internal degradation (23).

Table 2. Agents aggressive to inorganic materials

1. Chemical agents

1.1 Waters : soft water, sea water, rain
1.2 Acids : carbonic acid, hydrochloric acid, sulfuric acid, humic acid, lactic acid, tannic acid
1.3 Bases : soda, potash, urea, amines
1.4 Salts : chlorides, sulfates, nitrates, acetates, deicing salts, sea water
1.5 Alcohols
1.6 Oil : vegetable or animal

2. Biological agents

2.1 Bacteria
2.2 Algae, seaweed, shellfish
2.3 Insects, birds

3. Physical agents

3.1 Radiation: radioactive materials, solar radiation
3.2 Electricity: lighting
3.3 Magnetism and electromagnetism
3.4 Thermal agents: heat, fire, frost

4. Mechanical agents

4.1 Gravitation: loads
4.2 Deformations: shrinkage, creep, crystal pressure, expansions
4.3 Kinetics: vibrations, explosions, wind, shocks.

3.1 Degradation Mechanisms

3.1.1 Diffusion of chloride ions. The diffusion of chloride ions, occurring in concrete structures exposed to sea water or deicing salts, has a detrimental effect on reinforcing steel. So it is important to be able to evaluate the depth of the chloride penetration in concrete and to know what parameters of the building material can influence the diffusion of Cl^- ions to the cement paste-metal interface. Page, Short and El Tarras (22) have determined the coefficient of Cl^- ions diffusion through different types of cement paste with various cements (OPC, slag cement, fly ash cement) and various water/cement ratios. They found that the Cl^- diffusivity is strongly related to the pore structure and, more particularly, to the capillary porosity (pores larger than 1000 μm), which constitute continous channels in OPC with W/C = 0.6. The activation energy for diffusion of Cl^- ions calculated from the Arrhenius law is considerably higher in OPC pastes than in a normal electrolyte solution. So there is a limiting effect which may be due to some interactions as observed in clay-water systems. Blended cements with less capillary pores are more efficient than Portland cements in the protection of reinforcing bars in concrete structures (24).

3.1.2 Ettringite expansion. The mechanism of ettringite expansion in sea water or sulfate soils is not yet well understood. Several theories related to the formation of $C_3A{\cdot}3CaSO_4{\cdot}31H_2O$ have appeared recently. Some are contradictory in accounting for the following factors:

- cryśtal growth
- formation of colloidal ettringite
- water absorption
- increase of volume
- simultaneous processes
- supersaturation of the liquid interstitial phase.

All have tried to explain that the expansion is not always proportional to the amount of ettringite formed (Fig. 3). Nikitina et al (25) have shown that the supersaturation of the solution in lime, alumina and sulfate is a major factor in expansion phenomena. In an expansive cement, such as cement B (Fig. 3), there is no alumina in solution. Ettringite is formed at the surface of calcium aluminate particles leading to a localized increase of volume. This ettringite is often crystallized and has been considered as topochemically formed by some scientists, but it would be better to emphasize a local supersaturation of the solution and an absence of mass transfer to the liquid phase. On the contrary, in the stable alumina cement A (Fig 3), ettringite precipitated from a supersatured solution of alumina, lime and sulfate and appeared as well formed fibers which filled the pores without stresses. Calculated crystal growth pressures are considerably higher in expansive cements than in stable cements (23).

An example of the two types of ettringite, expansive and non-expansive, is given in the figure 4. If the ettringite, formed through solution and crystallized in the pores of the cement paste, is not expansive, more research is needed to explain the phenomenon.

Table 3. Example of detrimental process and detrimental effect of chemical attack for concrete (21) 1/

Chemicals	Process	Possible Effects
Monosodiumphosphate, NaH_2PO_4	0	0
Mineral oil	X	0ns
Diluted calcium chloride solution, $CaCl_2$	0e	0lrs
Strong calcium chloride solution, $CaCl_2$	CE	LR
Diluted sulphuric acid, H_2SO_4	Ac	Mk
Diluted hydrochloric acid, HCl	AE	MR
Water containing carbon dioxide	A	Mp
Carbon dioxide gas, CO_2	B	P
Ammonium chloride solution, NH_4Cl	BE	NPR
Calcium sulphate solution, $CaSO_4$	C	L
Magnesium sulphate solution, $MgSO_4$	CB	LN
Sodium hydroxide solution, NaOH	Db	Kn
Strong sodium chloride solution, NaCl	Ed	Rk
Hydrochloric acid fumes	E	r

Detrimental process

0. None
A. Dissolution of solid components
B. Transformation of solid components (ion exchange, polymorphic changes)
C. Formation of expansive compounds
D. Reaction with aggregates
E. Corrosion of reinforcement
X. Still unknown

Resulting detrimental effect

0. None
K. Deformation
L. Cracking
M. Loss of mass
N. Change in mechanical characteristics
P. Loss of alkalinity
R. Weakening of reinforcement
S. Change in color
X. Others

1/ Capital letters stand for distinct main processes and effects. Small letters stand for small, secondary or only possible processes or effects. Each classification begins with a capital letter. An 0 indicates that there is no distinct process or effect.

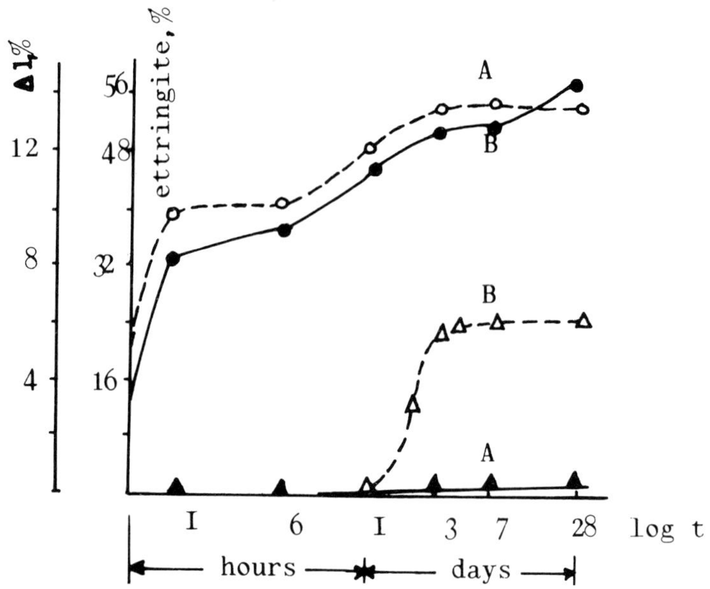

FIG 3. Amount of ettringite (O) and linear expansion (Δ) in two cement pastes (25).
A = alumina cement + 30% gypsum
B = alumina cement + 30% gypsum + 5% CaO

3.1.3 Alkali-aggregate reaction. Concretes made with aggregates containing amorphous silica (opal, chalcedony) or siliceous aggregates (gneiss, schist, sandstone, basalt, quartzite) or dolomitic limestone can be degraded by the formation of alkaline expansive silicate or by the loss in bond strength. Studies carried out over the last forty years have come to the conclusion that the degradation is caused by the simultaneous influence of factors such as:

- The alkali content in the cement and the alkali liberated by some aggregates such as feldspars or admixtures
- The content and type of unstable minerals in the natural aggregates: "pessimum content"
- The wet and drying cycling or
- The temperature.

Some points need to be cleared up in the near future i.e. (26):

- The mechanism of concrete expansion and concrete cracking (osmotic pressure or imbibition pressure)

- The transformation of the gel into viscous liquid which moves through the cement paste and fills up the pores (influence of Ca^{+2} ions)

- The expansion of concretes made with Portland cement containing only 0.4% Na_2O (source of alkalies).

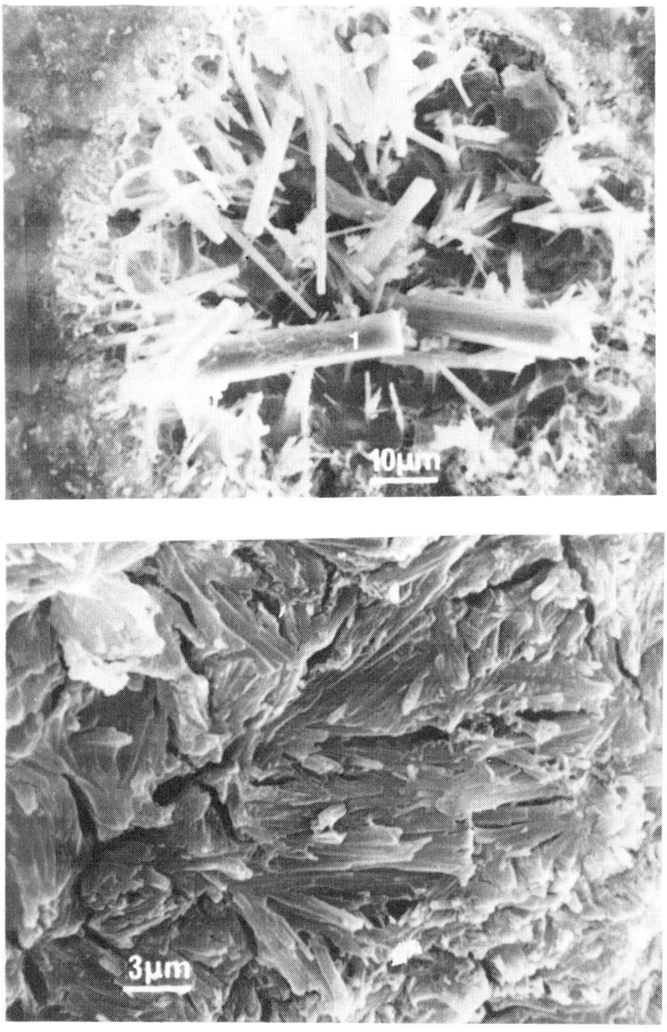

FIG 4. Crystallization of ettringite in cement pastes stored in sea water

a. non-expansive crystals in a pore
b. expansive ettringite poorly crystallized at the surface of aluminates.

- The occurrence of visible deterioration in bridges, dams, roads and dikes between five and thirty years after the placing of concrete (influence of shrinkage, temperature, humidity)
- The difficulty in repairing a damaged concrete as cracks reappear beneath resin or paint coatings (Fig. 5) and
- The failure of some accelerated tests.

3.2 Accelerated Tests

Accelerated tests must induce, in the laboratory and in a short period of time, corrosion processes which in natural environments can take many years. The chemical reactions are accelerated by either increasing the temperature or increasing the concentration of aggressive agents. Mechanical strength or linear expansion is measured on mortar or concrete samples of reduced dimensions in order to assess the effect of the exposure environment.

Some tests are too severe. In the AAR case, they have eliminated materials, the resistance of which would have been favorable over the long term.

In other cases, they are unsatisfactory because they over-simplify the mechanism of degradation. Regarding AAR, the optimum test seems to depend on the type of aggregate (silica, volcanic glass, carbonate rock, quartzites and greywacke). The C 289-71 chemical test, measuring the dissolved silica, has given either pessimistic results for English flint aggregates or, more often, optimistic results, as for deleterious South African shales. The C 227-71 physical mortar bar test can be difficult or misleading due to unexplained contraction or only progressive expansions. The C 586-69 rock cylinder test works differently for dolomites and active carbonate rocks from different origins and a poor correlation has been found between the expansion of rock cylinders and concrete prisms. Modification of ASTM C 289 and C 227 tests has been presented by Oberholster (27) at the last international meeting in 1983. Petrographic examinations of aggregates should never be omitted although interpretation requires experienced scientists. So, accelerated tests are useful, but their results must be interpreted carefully.

3.3 Preventive Measures

The need for preventive measures is a function of the degree of aggressivity of the environment but, in all cases, a well made and well designed concrete with high impermeability rarely deteriorates rapidly. Measures for the mix design of concrete take into account the type of cement, the water/cement ratio and the cement content of concrete. In some cases, an additional protection is necessary.

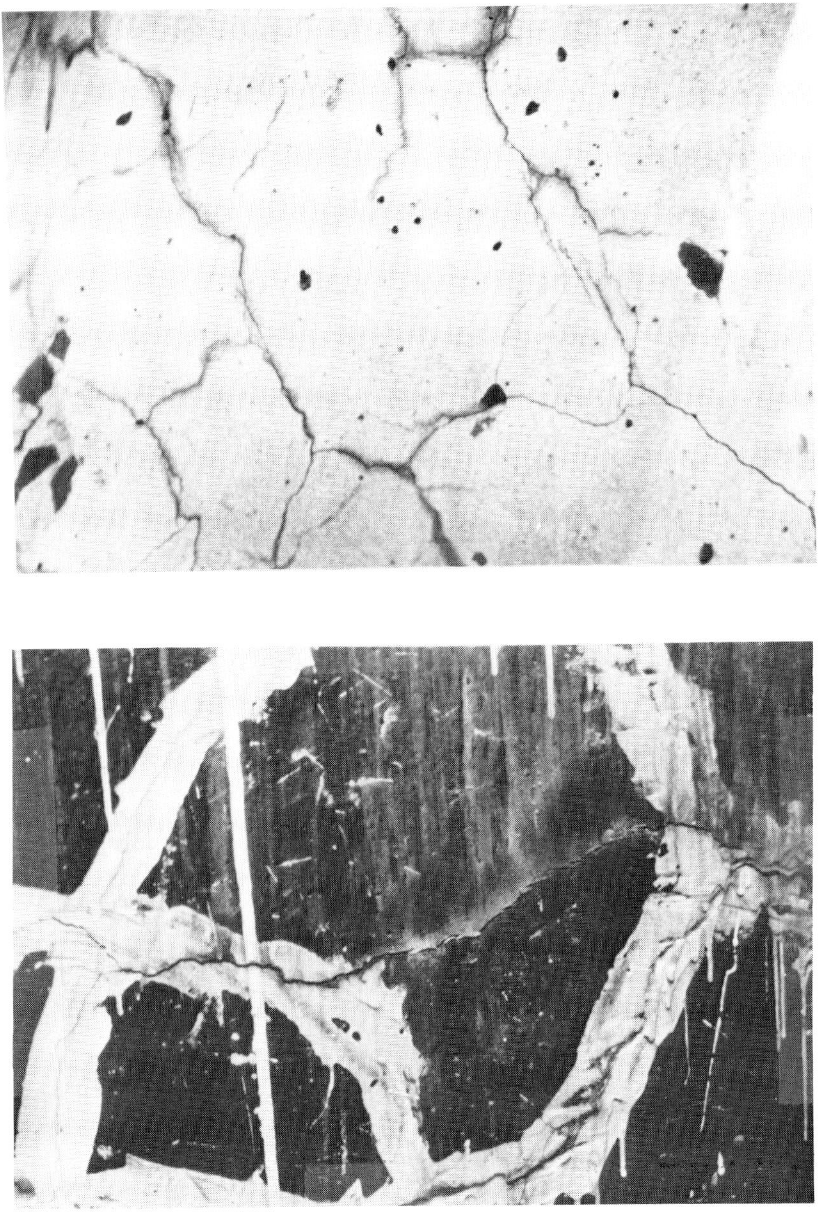

FIG 5. Concrete deteriorated by alkali-aggregate reaction
a) typical map cracking in a humid environment
b) reappearance of cracks after treatment (resin and paint)

3.3.1 Choice of cement. Different types of cement have been selected for special uses, i.e. structures in sea water, sulfate soils, carbonated atmospheres or acid environments. The basis for cement selection results from long experience. The case of the alkali-aggregate reaction is a relatively recent one. In the presence of potentially reactive aggregates:

- The ASTM C 33 standard recommends Portland cements with a total amount of alkalies less than 0.6% Na_2O
- The German recommendations concerning blast furnance cements are (28):
  - slag content ⩾ 50% for $-Na_2O$ eq ⩽ 0.9%
  - slag content ⩾ 65% for $-Na_2O$ eq ⩽ 2%
- The British rule for reactive siliceous aggregate in concrete is (29):
  $-C\ (Na_2O + 0.658\ K_2O) \leqslant 4\ kg/m^3$

where C is the cement content in kg/m^3

- The South-Africa rule for siliceous aggregates (Malmesbury Shales) considers (27):
  - between 3.8 and 1.8 kg of Na_2O eq per m^3 of concrete: the reactivity of aggregate is tested and the active alkali content of the concrete is estimated
  - above 3.8 kg Na_2O: deleterious expansion always occurs
  - below 1.8 kg Na_2O: no expansion occurs.

3.3.2 Mixing water. Tap water is the best mixing water but in some cases river water, sea water or industrial water is the only possibility on the work site. These latter three waters must be analyzed before use. The amount of dissolved salts and solids in suspension has been respectively limited to 36 g/l and 5 g/l for plain concretes. When sea water is used, the cement hydration is accelerated due to the presence of chlorides. In hot countries, a retarder is sometimes necessary. Mixing prestressed concrete with sea water is not permitted.

3.3.3 Aggregates. The granularity, texture and mineralogical composition of aggregates play a role in the resultant mechanical properties and behavior of concrete. The amount of distribution of opal, calcedoine and feldspar must be known if the concrete is used in a humid environment. Several aggregates are known to react with alkalies. These have been compiled by various countries.

Clay particles or organic materials can retard the setting of cement or lead to a swelling by water adsorption.

3.3.4 Making and placing of concrete. The compactness of the material is a main factor concerning concrete durability. The water/cement ratio has to be lowered in aggressive environments, using superplasticizers, for example.

In order to prevent the diffusion of aggressive agents, the cement must have reacted sufficiently, thus becoming resistant before its contact with corrosive agents. However, a rapidly curing cement can microcrack due to shrinkage. So a cement that is slowly reactive with water and which liberates a low heat of hydration has been preferred to an early strength cement.

3.3.5 Secondary protection. In highly aggressive environments, protection of the concrete surface is necessary. In very aggressive environments, resin impregnations can constitute a supplementary protection.

In the case of possible biological deterioration, the concrete surface can be protected by a layer of clay containing sulfur reducing bacteria. This represents a temporary protection and cannot be used for reinforced concrete. Protection against bacteria may be improved if limestone is added to the clay.

4 CONCLUSION

This paper has been mostly related to the durability of concrete, a composite inorganic building material. The service life of a given concrete structure depends on its:

- <u>environment</u>: climate and microclimates, polluted atmospheres and acid rains in industrial regions, aggressive waters and soils, moisture transfer in humid countries or evaporation and transpiration in hot dry countries

- <u>mechanical performance</u>: strength, shrinkage and cracking which are all related to the intrinsic resistance of constituents and the bond between them

- <u>physical and chemical resistance</u> functions of the constituents, i.e. their own chemical resistance to aggressive media but also their behaviour in complex systems where secondary reactions occur due to a preliminary deterioration. For example, massive ettringite is always found close to aggregates which have first reacted with alkalies.

The prediction of service life is closely related to the knowledge of degradation mechanisms, accelerated tests and preventive measures:

- Degradation mechanisms involve many parameters, the action of which is not yet completely elucidated as, for example, the amount of reactive particles or inclusions in siliceous aggregates, called pessimum content in AAR.

- Accelerated tests are not always suitable because they may be over-simplified. The ASTM tests regarding AAR established for sands do not work well with all siliceous aggregates.

- Preventive measures follow the knowledge of degradation mechanisms but must also take into account new products, such as cements, admixtures and reinforcements which result from either energy saving efforts or performance optimization.

The barriers to the prediction of service life of inorganic building materials will decrease when there is a closer collaboration between civil engineers and materials scientists on the work site for:

- a better characterization of the environment
- a precice understanding of the possible degradations
- a suitable choice of materials and design and
- a modelling of materials and structures behaviour.

Examples of such collaboration exist. They have been successful but there are still only a few.

5 REFERENCES

1. Wittman, F. H., Modelling of Concrete Behavior. Contemporary European Concrete Research. The Swedish Concrete Inst., 171-190, 1981.
2. Mindess, S., Mechanical Performance of Cementitious Systems. Structure and Performance of Cements. P. Barnes Ed. Applied Sci. Publishers. 319-364, 1984.
3. Roelfstra, P. and Wittmann, F. H., Modelling the Time - Dependent Behavior of Concrete. Transactions of the 6th SMIRT. Vol. L. Paper L 1/4, 1981.
4. Zaitvev, J. and Wittmann, F. H., Simulation of Crack Propagation and Failure of Concrete. Mat. Struc. 14 (83), 357, 1981.
5. Mihashi, H. and Izumi, M., Stochastic Theory for Concrete Fracture. Cem. Concr. Res. 7, 411-421, 1977.
6. Sereda, P. J., Feldman, R. F. and Ramachandran, V. S., Formation and Development of Structures in Hardened Cement Pastes. 7th Int. Congress on the Chemistry of Cements. Paris. Vol. I, VI - 1/3, 1980.
7. Wittman, F. H., The Structure of Hardened Cement Pastes. A Basis for a Better Understanding of the Materials Properties. Proc. Conf. on Hydraulic Cement Pastes. Their Structure and Properties. 96-117, 1976.
8. Copeland, L. E. and Verbeck, G. J., Structure and Properties of Hardened Cement Pastes. A Principal Paper. 6th Int. Congress on the Chemistry of Cement. Moscow, 1974.
9. Powers, T. C. and Brownyard, T. L., Studies of the Physical Poperties of the Hardened Portland Cement Paste. Portland Cement Association. Res. Bull 22, 1947.
10. Kendall, K., Howard, A. J. and Birchall, J. D., The relation between Porosity, Microstructue and Strength and the Approach to Advanced Cement - Based Materials. Phil. Trans. R. Soc. London. A 310, 139-153, 1983.
11. Powers, T. C. and Helmuth, R. A., Theory of Volume Changes in Hardened Portland Cement Paste during Freezing. Proc. Highway Res. Board. 32, 1953.
12. Litvan, G. G., Freeze-Thaw Durability of Porous Building Materials, ASTM. Spec. Tech. Publ. 691, 1981.
13. Fagerlund, G., Testing of Frost Resistance. Int. Colloquium on Frost Resistance of Concrete. Vienna, 1980.
14. Powers, T. C., The Air Requirement of Frost Resistant Concrete. Proc. Highway Res. Board. 19, 1949.
15. Fagerlund, G., The Critical Degree of Saturation Method of Assessing the Freeze-Thaw Resistance of Concrete. Mat. Struct. Vol. 10, No. 58, 1977.
16. Pigeon, M., La Microstructure et la Résistance au Gel des Ciments et Bétons. Thèse Doc. Ing. Paris. VI - Juin 1984.

17. Litvan, G. G., Frost Action in Porous Systems. Int. Sem. on Durability of Concretes and Stones. St. Rémy-lès-Chevreuse, France, 95-108, 1981.
18. Fagerlund, G., Service Life of Concrete Structures. Ibid Ref. (1), 265-285, 1981.
19. Houst, Y., Roelfstra, P. E. and Wittmann, F. H., A Model to Predict Service Life of Concrete Structures. Int. Colloquium on Materials Science and Restoration. Esslingen. W. Germany, 181-186, 1983.
20. Vénuat, M. et Alexandre J., De la Carbonation du Béton. Rev. Mat. Constr. 683, 421-427 (1968), 639, 469-481 (1968), 640, 5-22 (1969).
21. Regourd, M., Resistance of Concrete to Chemical Attacks. 32 RCA Rilem. Activity Report. Mat. Struct., 14, 80, 109-137, 1981.
22. Page, C. L., Short, N. R. and El Tarras, A., Diffusion of Chloride Ions in Hardened Cement Pastes. Cem. Concr. Res. 11, 3, 395-406, 1981.
23. Regourd, M., Chemical Corrosion of Mineral Building Materials. Ibid Ref. 19. To appear in Bautenschutz + Bausanierung (1984).
24. Smolczyk, H. G., Structure and Characterization of Slags. Principal Paper. Theme III/1. 7th Int. Congress on the Chemistry of Cement. Paris. Vol. I, III - 1/3, 1980.
25. Nikitina, L. V., Krassilnikov, K. G. and Lapchina, A. I., La Nature Physico -chimique des Autocontraintes des Ciments Expansifs. Ibid Ref. 23. Vol III, V. 39-44, 1980.
26. Regourd, M., Durability - Physico-chemical and Biological Processes Related to Concrete. Int. Workshop on Durability of Concrete Structures. Copenhagen (1983) Paper 1.2.
27. Oberholster, R. E., Alkali Reactivity of Siliceous Rock Aggregates: Diagnosis of the Reaction, Testing of Cement and Aggregate and Prescription of Preventive Measures. 6th Int. Conf. Alkalies In Concrete, Copenhagen, 419-433, 1983.
28. Bakker, R. F. M., About the Cause of the Resistance of Blast-furnance Cement Concrete to the Alkali-Silica Reaction, 5th Int. Conf. on Alkali Reaction in Concrete. Cape Town, South Africa. S 252/29, 1981.
29. Hobbs, D. W., Expansion of Concrete Due to Alkali-Silica Reaction. The Structural Engineer. Cement and Concrete Association Reprint, 1984.

BARRIERS TO THE PREDICTION OF SERVICE LIFE OF METALLIC MATERIALS

Christopher L Page

University of Aston, Birmingham, UK

ABSTRACT

The degradation of metallic materials in buildings and other structures is frequently attributable to electrochemical corrosion processes. In this paper, general problems of corrosion prediction are first considered and it is shown that information derived from accelerated exposure tests or from investigations of electrochemical properties of idealised systems is often an inadequate guide to the behaviour of metallic components in real corrosive environments. Specific cases involving the durability of steel reinforcing bars in concrete and the behaviour of zinc-coated steel components in buildings are then discussed. These examples are chosen because they illustrate the difficulties involved in predicting service lives of metals that are exposed to complex and often ill-characterised environments.of the sorts that may be encountered in constructional applications.

1 INTRODUCTION

The range of metals and alloys employed by the construction industry is very wide as are the fields of application of these materials. During the last half-century innovations in building practices and competition from alternative materials have promoted numerous changes in the constructional uses of metals. As a consequence, it has become frequently necessary to predict service-lives of components for which there is no direct experience of long-term durability. The problem is compounded by the lengthy periods, typically in excess of fifty years, for which such components are expected to function adequately and by the serious consequences often implied by premature deterioration, in terms of safety, costs of remedial work etc.

To form a general view of service-life prediction, it is necessary to examine the effects and possible interactions of a large number of degradative agencies of the sorts classified in recent standard recommendations for building materials and components (1,2). This approach is as valid in the case of metals as for other materials since failures may be caused by processes such as corrosion, fatigue, fracture or combinations thereof, being subject to the influence of numerous environmental and stress-related factors.

For a large majority of practical instances, however, the degradation of metals in constructional applications is associated, at least in part, with some form of corrosion stimulated by contact between the metal and an aqueous medium. In the present contribution, consideration will be restricted to processes of this kind and an attempt will be made to exemplify the main barriers to prediction of corrosion performance.

2 GENERAL PROBLEMS OF CORROSION PREDICTION

Corrosion prediction is important in all fields of engineering where metals are used, the chemical industries, in particular, being faced with serious economic penalties if component failures cause processes to be shut down at times when regular maintenance had not been anticipated. There has therefore been substantial research and development into methods of predicting corrosion behaviour and it is worthwhile to examine some general difficulties in this area before considering specific cases of interest in the field of construction.

Discussion of corrosion problems rests on the now well-established principle that the mechanisms by which metals interact with aqueous media are electrochemical in nature. This largely explains why the corrosion exhibited under different circumstances can take such a variety of forms; whilst the total rates of anodic and cathodic processes occurring in a given system must be balanced, the distribution of anodic and cathodic regions may vary greatly, giving rise to the following possibilities:

a) general corrosion, implying a random distribution of anodic sites and hence uniform attack of the exposed surface;

b) galvanic corrosion, involving preferential anodic dissolution of the more reactive component of a bimetallic couple or segregated alloy;

c) localised corrosion, embracing a number of processes in which occluded cells are formed and where breakdown of protective (passive) films may often be involved, eg. crevice corrosion, pitting, stress corrosion cracking and corrosion fatigue.

Service life prediction is thus frequently complicated by the need to take account not only of factors that influence the intensity of a given form of attack but also of critical conditions that may signify a sudden switch in the type of corrosion observed.

Approaches that are often followed in relation to corrosion prediction make use of various accelerated testing procedures (3). Those commonly employed in industry for purposes of routine quality control and acceptance testing tend to be purely empirical methods which depend on increasing the severity of the corrosive environment by controlling variables such as temperature or concentrations of aggressive reactants. Such techniques can never provide more than a simple guide to the relative susceptibilities of different materials to particular forms of corrosion and they may prove to be misleading if used uncritically since it is only too easy to distort normal corrosion reactions.

More scientific attempts to predict corrosion behaviour are based on study of the properties of model electrochemical systems, which have features in common with the corroding systems under consideration. Thermodynamic data may be used to establish the conditions under which particular reactions are possible or impossible and thus to define the domains of potential and electrolyte composition where, for example, general corrosion processes or film-forming reactions that may induce passivation of the metal are likely (4). Studies of anodic reaction kinetics under conditions of potentiostatic polarisation may be used to provide supporting evidence of these forms of behaviour and to yield additional information regarding the domains in which aggressive anions may break down passive films to cause pitting (5), see Fig. 1. For well-characterised corroding systems involving bulk electrolytes, in which the forms of the constituent anodic and cathodic polarisation curves may be accurately determined, it is a simple matter to estimate the mixed potential and average corrosion rate by construction of a standard Evans diagram. Since the forms of polarisation curves may often be represented by equations containing parameters that are expressible in terms of variables such as temperature, reactant concentrations, diffusion rates etc. (6), it is possible, at least in principle, to predict the effects of changes in these variables on corrosion kinetics. This approach has been used recently in construction of a computer-based model of the behaviour of a system for which extensive polarisation data are available, viz. stainless

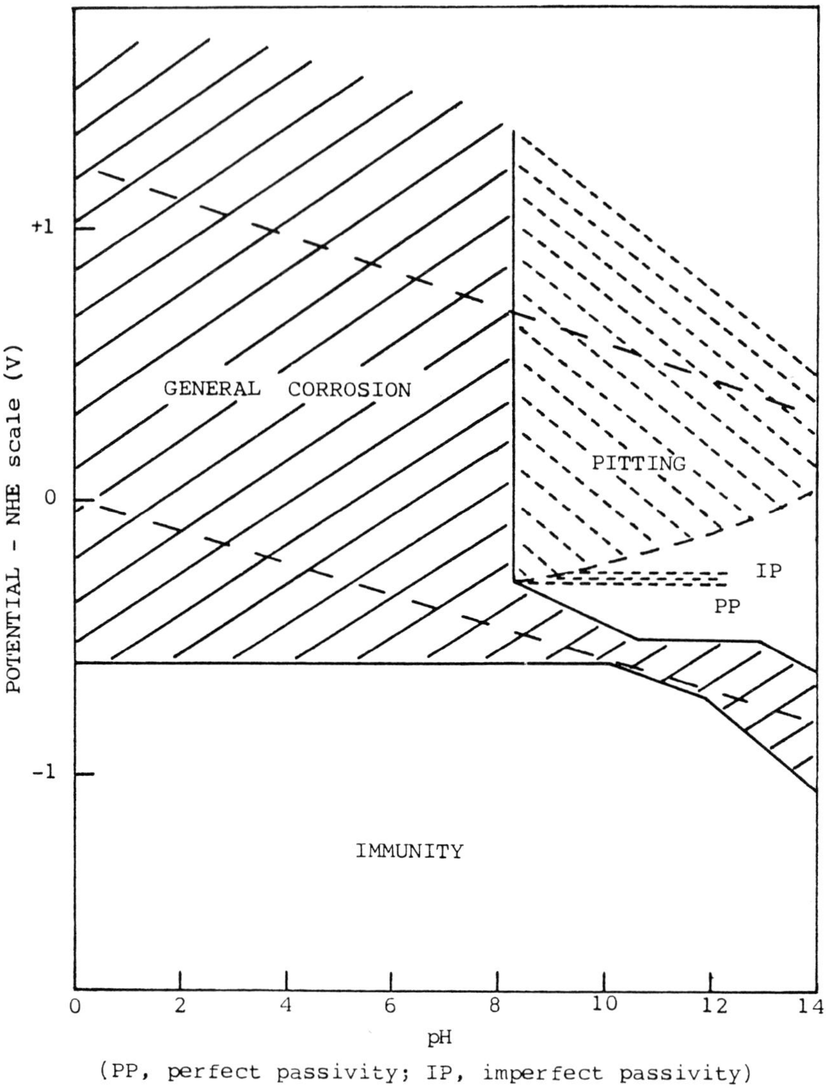

Figure 1 Potential/pH diagram for Fe in 1 M Cl^- solutions

steel (type 304L) in sulphuric acid solutions (7).

Unfortunately, however, there are a number of practical objections to the use of such procedures as anything more than a qualitative guide to corrosion prediction for the majority of real systems. Thus, obvious difficulties are faced in assessing the effects of surface condition, physical form and configuration (crevices etc.) of metals in service on the basis of information derived from standard laboratory measurements. In addition, actual service environments are often highly complex and difficult to characterise. Components in buildings, for example, may be exposed under various circumstances to contact with bulk aqueous solutions, thin films of moisture formed by condensation from the atmosphere or solutions contained within the pores of other materials such as concrete and timber. The chemical compositions of many of these electrolytes are liable to vary with time and their physical forms often militate against the use of standard techniques for acquiring electrochemical data such as polarisation curves.

From the forgoing considerations, it is clear that study of the behaviour of idealised systems often provides little help in resolving matters of service-life prediction. It is necessary rather to approach such problems by procedures that take account of the specific complexities of the particular corroding systems under consideration. The first essential step is therefore to obtain an accurate description of the process or processes leading to failure. Attempts may then be made to model these processes with a view to identifying critical rate-determining parameters, the values of which may be estimated either from records of service under related conditions or from carefully designed exposure tests. The latter should simulate the actual service environment as closely as is practicable and it may often be advantageous to employ non-destructive electrochemical methods to monitor the time-dependence of corrosion rates. Significant recent progress has been made in developing such techniques for measuring rates of atmospheric corrosion (8) and for use in other complex corrosive media that will be referred to later in this paper.

If such an approach is followed then it should at least serve to identify areas where current understanding or existing data need to be supplemented to permit more accurate assessments of service-life to be made. In the present state of knowledge, as illustrated by the specific cases to be discussed in the following section, this may be considered a useful advance.

3 SPECIFIC CASES

3.1 Steel Reinforcement in Concrete

A good illustration of the complexity of practical corrosion problems is provided by the behaviour of steel reinforcing bars in concrete, the subject of recent reviews (9,10). There is currently much interest in this topic because of growing evidence of premature deterioration of reinforced concrete structures in many countries and a report, intended to serve as a comprehensive basis for modelling, is being prepared by Technical Committee 60-CSC of RILEM.

Concrete is a porous material in which an alkaline aqueous phase is normally present since NaOH, KOH and $Ca(OH)_2$ are formed during the hydration of Portland cements. With the development of techniques for sampling and analysis of this pore solution, it has been shown that pH values in the range 13 to 14 are typical (11) and embedded steel therefore displays passivity under most circumstances (4). In this condition, the corrosion rate is negligibly small but there are two common agencies that may cause breakdown of passivity. These are acidic gases such as atmospheric CO_2, which may carbonate concrete depressing the pH value of its pore solution towards neutral (non-passivating) levels, and chloride salts, derived from contaminated mix materials or from contact with saline constituents of certain service environments, which may induce depassivation (pitting) even at high pH values, as depicted in Fig. 1. In either case the formation of corrosion products which are expansive generates tensile stresses that may cause the concrete to crack or spall and this represents the normal mode of failure when there is a need for remedial treatment. Circumstances occasionally arise in which significant loss of cross-section of the embedded steel may endanger the structural strength of a member before cracking of the concrete becomes obvious, but these are more relevant for prestressed concrete structures.

For reinforced concrete components, therefore, parameters that need to be estimated for purposes of service-life prediction are generally as follows:

a) the initiation time (t_o) during which the steel remains passive;

b) the subsequent rate of corrosion as a function of time, from which an average rate (i_{corr}) may be estimated;

c) the maximum tolerable average amount of corrosion per unit area of the steel (Q_{max}) corresponding to the condition of failure.

The corrosion propagation time (t_1) is equal to (Q_{max}/i_{corr}) and the predicted service life is ($t_o + t_1$).

It is therefore necessary to consider how reliably these components may be estimated as a means of identifying current barriers to service life prediction for this system.

a) Initiation Time (t_o)

When carbonation is the depassivating agency, t_o is the time required for the surface layer of neutralised material to penetrate to the level of embedded steel. The process is governed by the rate of inward diffusion of CO_2 to a reaction zone where $CaCO_3$ and other products are formed.

For the simplest case of unidirectional diffusion into concrete of uniform stable pore structure and moisture content (implying no variation of the diffusion coefficient with time, t, or depth, x) from an environment of constant temperature, relative humidity and CO_2 concentration, C, the rate of advance of the carbonation front would be:

$$\frac{dx}{dt} = \frac{DC}{Ax} \qquad (1)$$

(where A represents the quantity of alkaline substances reacting per unit volume of concrete).

On integration this yields a parabolic relationship:

$$x^2 = \frac{2DCt}{A} \qquad (2)$$

The assumptions involved in this simple treatment clearly cannot apply in reality to exposed concrete which is subject to changes in moisture condition etc. and more rigorous analyses, which take account of variation of D with depth, are possible (12). Nevertheless, empirical data on carbonation can be approximately fitted to parabolic relationships (9) and equation 2 displays several of the main variables that need to be considered. Thus, the value of D is dependent on temperature, moisture content and concrete quality (water/cement ratio, compaction and curing); the value of A is related to cement content and cement composition; C is dependent on the level of atmospheric pollution.

Most of the studies of carbonation kinetics that have been made relate to Portland cement concretes covering a normal range of mix compositions, exposed in sheltered or wet temperate locations. There is therefore a considerable body of published data which may be used to estimate values of t_o corresponding to specified depths of cover under these circumstances (9), see Fig. 2. Information related to hot climates, where difficulties of production and curing of concrete may exist, is limited, however, and there is a need for further research in this area (13).

When corrosion is caused by chloride ions, as opposed to carbonation, values of t_o may range from zero for concrete with a high level of initial salt contamination to a substantial fraction of the total service life for structures exposed to environments containing seawater, deicing salt etc. A major difficulty in predicting t_o is that the threshold concentration of chloride required to depassivate steel in concrete depends on a large number of factors which have been discussed in detail elsewhere (10). Several of these are currently under investigation as there is a need to establish more accurate guidance rules.

If a particular chloride threshold is assumed, however, such as 0.4% chloride by weight of cement for concrete manufactured from Ordinary Portland cement (10), then values of t_o for structures exposed to chloride-containing environments may sometimes be estimated. In cases where concrete is permanently near to being saturated and chloride ingress is governed by diffusion, penetration profiles may be calculated from standard solutions of Fick's second law. Thus, equation 3, which applies to a semi-infinite medium exposed to a constant surface concentration of the diffusing species (C_s), has been used successfully to model the penetration of chloride ions into 15-20 year old concrete marine structures in locations up to 15 metres above water-level (14):

$$C_s - C_x = (C_s - C_o) \operatorname{erf} \frac{x}{2\sqrt{Dt}} \qquad (3)$$

where, C_x = concentration at distance, x, from the exposed surface after time, t.

C_o = original concentration

D = diffusion coefficient

The diffusion coefficient varies with concrete quality, cement type and temperature (15) and there is a need for further research to determine the effects of these factors, particularly for concretes of low intended permeability.

For cases where concrete is exposed to alternate treatment with salt water and drying, streaming effects may convey chloride ions more rapidly than diffusion would do. Kinetics do not appear to have been studied in a systematic way and it is therefore

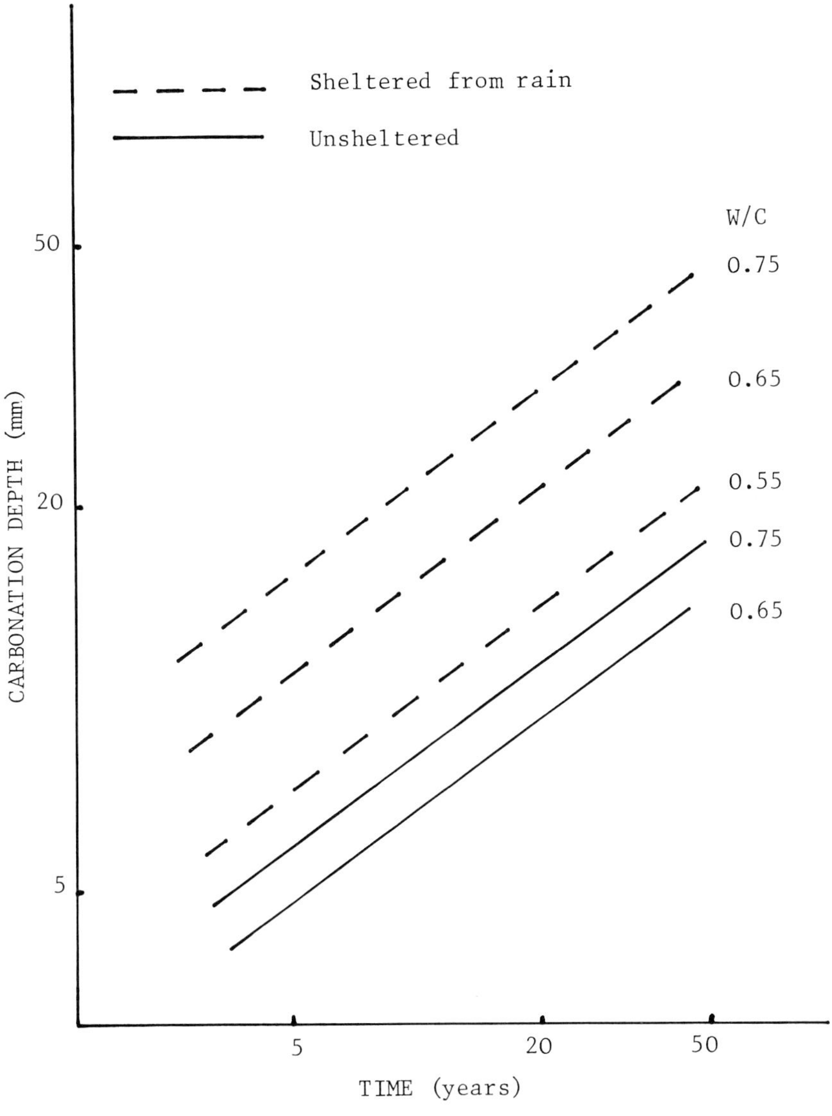

Figure 2 Typical depths of carbonation recorded for Portland cement concretes of various water : cement ratios (W/C).

impossible to estimate penetration profiles under varying conditions of drying frequency and severity. Practical experience of the behaviour of reinforcing bars in structures exposed to regular treatment with deicing salt, however, suggests that it is extremely difficult to extend the corrosion initiation period (t_o) beyond about five years when structural concretes of normal quality and cover thickness are used.

b) Corrosion Rates (i_{corr})

Data for rates of corrosion of depassivated steel in concrete under various conditions of exposure have proved somewhat difficult to obtain, but significant progress has been made during the last few years. This has been due mainly to the successful application of non-destructive electrochemical techniques of corrosion rate determination, based on the linear polarisation approach devised by Stern and coworkers (16, 17). Methods of this sort, when used under appropriate conditions (i.e. with IR compensation, suitable potentiodynamic scan-rate etc.), have been shown to yield reliable data that are well correlated with corrosion rates obtained by destructive gravimetric measurements (18). They may therefore be used to assess effects, such as those arising from changes in the moisture condition of specimens which have been carbonated or contaminated with chloride ions (19), see Fig. 3, or to evaluate the influence of chloride penetration (20), see Fig. 4.

The practical importance of such data is that they provide a basis for predicting whether, for example, the effects of carbonation or chloride contamination would be likely to stimulate significant rates of corrosion under known atmospheric conditions. There is an urgent need for further detailed information of this sort to be obtained since rational assessment of the risks of premature corrosion damage in many structures is dependent on it.

Whilst electrochemical methods of the sort discussed have an important role, they suffer the disadvantage of requiring the metal studied to be in the form of a reasonably small electrode that does not form part of the actual network of reinforcing bars in a structure. This limits their suitability in cases where macroscopic corrosion cells are likely to be important, eg. in concrete of modest resistivity with large variations in chloride content. Recent attempts to develop a modified form of the linear polarisation technique, which is intended to permit measurements of corrosion rates to be made for segments of a continuous network of steel bars, are therefore of interest in this context though further development is necessary (21).

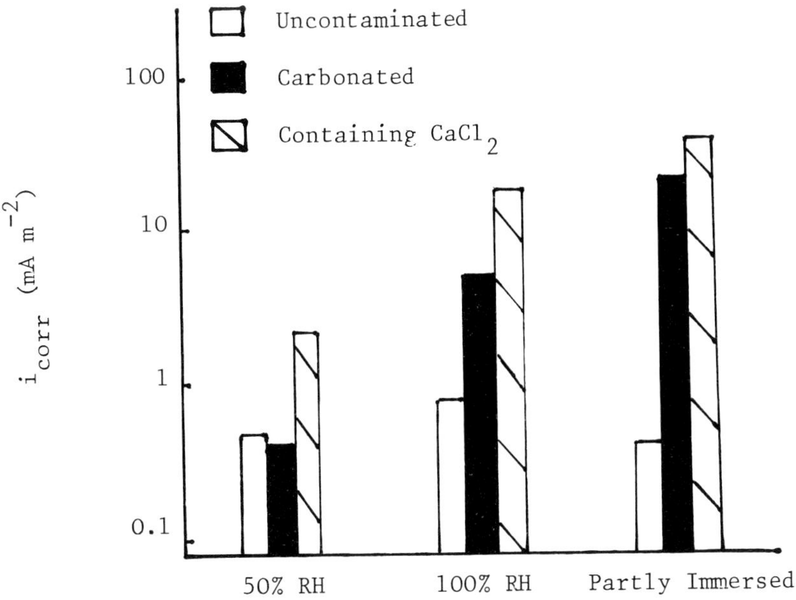

Figure 3 Effects of exposure condition and contamination on typical corrosion rates of steel in mortars

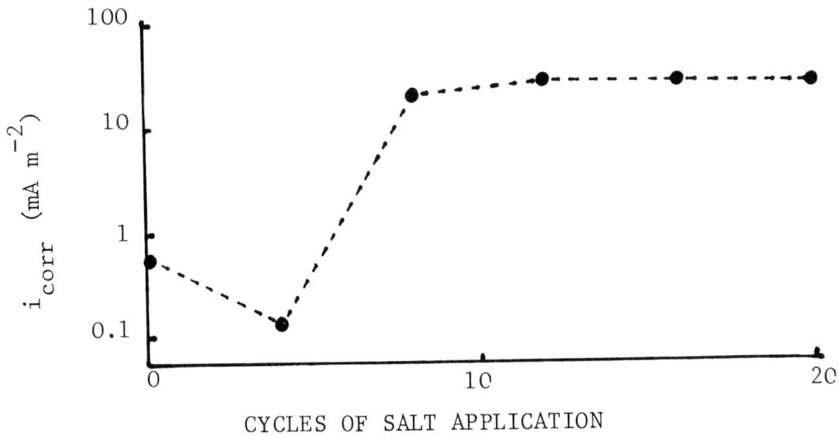

Figure 4 Effects of chloride penetration into concrete exposed to treatment with deicing salt on corrosion rate of embedded steel

c) Maximum Tolerable Corrosion (Q_{max})

Definition of the maximum amount of corrosion that can be tolerated poses a number of problems, the first of which is to decide the degree of cracking that constitutes an appropriate limit to service life. Clearly this will depend on the sort of structure concerned and its location.

If, for the sake of argument, the time taken to form visible cracks at the surface of the concrete is specified, there is a paucity of published information concerning Q_{max} under different circumstances. Values are likely to depend on variables such as bar diameter, cover thickness and concrete quality as well as being related to the nature and mobility of the corrosion products formed.

According to some recent estimates, average thicknesses of corrosion product formed prior to cracking of either carbonated or chloride-contaminated concretes with reinforcement at cover thicknesses of 20 mm or more are often in the range of a few hundred µm to 1 mm (9). Other reports, however, have suggested that smaller quantities of corrosion, amounting to average thicknesses of Fe consumption of 10-50 µm, can be sufficient to form cracks in concrete of water/cement ratio, 0.5, containing bars of 10 mm diameter at a depth of 25 mm (22). Many factors could account for these differences and it is clear that further detailed study of the variables affecting Q_{max} is required.

3.2 Galvanised Steel Components

Another area in which problems of corrosion prediction have recently provoked concern is associated with the performance of zinc-coated steel components of various sorts used in building. Zinc coatings, applied mainly by the process of hot-dip galvanising, have a long history of use as a means of providing corrosion protection to iron and steel components and it is therefore surprising that questions of appropriate coating thickness to achieve a required service-life still give rise to difficulties.

In the UK, however, recent problems involving unexpectedly rapid deterioration of galvanised steel wall ties, which provide connection between the inner and outer leaves of cavity-wall constructions, have been encountered (23). As a result of these problems, it has been necessary to make substantial revisions to the British Standard specification for these components, increasing minimum recommended coating thicknesses (expressed as zinc weights per unit area) to 940 g/m^2 (24). There has also been speculation about the adequacy of currently recommended coating thicknesses of certain other galvanised steel products and this prompted the British Board of Agrément to commission a major review of the subject, which has been recently published (25). This report merits careful study as it highlights a number of the main difficulties associated with service-life prediction for metallic components in buildings.

In the case of galvanised steel wall ties, the essential cause of failure to predict coating performance can be traced to ignorance of the nature and severity of the corrosive environment to which metals are exposed in a cavity-wall. Until recent investigations proved otherwise, it was apparently considered that the portion of a metal wall-tie exposed to the greatest risk of corrosion was the section bridging the atmospheric zone within the cavity itself. This zone, if assumed to represent a generally dry interior atmosphere, might reasonably have been expected to sustain average zinc corrosion rates of less than 5 $g/m^2/year$ (25) so the minimum specified coating thicknesses prior to the 1981 amendment of the relevant British Standard (i.e. 260 g/m^2 for wire ties) would have been considered adequate for the required service-life of 60 years or more.

In practice, however, recent investigations of the condition of wall-ties in structures have shown that the regions exposed to the greatest risk of corrosion are the sections of metal embedded in the mortar beds of the outer leaf of a cavity-wall. Many factors can affect the severity of attack in this zone, including the frequency and intensity of exposure to driving rain, the presence of aggressive contaminants in the mortar etc. Average rates of consumption of zinc, however, have typically been found to lie within the range 10-20 $g/m^2/year$ (23). These figures, obtained from direct measurements on ties removed from buildings, are in reasonable agreement with corrosion rates determined from linear polarisation studies of galvanised steel electrodes embedded in mortars, which were subjected to carbonation and then exposed to an atmosphere at 100% relative humidity or to partial immersion in water (26). Thus, since it is known that the mortar beds in brickwork carbonate rapidly and may become soaked for substantial periods, the observed failures of wall-ties might have been predicted by means of non-destructive electrochemical monitoring.

Apart from its obvious implications with respect to other galvanised steel components, such as joist hangers and lintels, that are exposed in similar environments to those in which wall-ties are placed, this recent spate of failures has raised wider questions about our current knowledge of certain features of the internal environment in buildings. In comparison with the extensive literature concerned with corrosion of galvanised steel in external atmospheres, very little appears to have been published about effects of exposure to cold, moist internal air within cavities, roof-spaces etc. where there is a danger of condensation. Specific forms of interaction between galvanised steel and certain roof insulant materials or timbers treated with inorganic salt flame retardants or preservatives based on copper-chrome-arsenate have been identified as potential stimulants of rapid corrosion. There is, however, a recognised need for more general investigation of the kinetics of corrosion of galvanised steel in internal situations and some recently initiated exposure tests, aimed at simulating the environments within well-ventilated lofts at marine, industrial and rural sites, are reported to be in progress (25).

4 CONCLUSIONS

The examples considered in this paper illustrate that the present state of knowledge with regard to service life prediction of metallic materials in constructional applications is far from adequate. In many instances the main difficulty encountered is lack of detailed information about the nature of the service environment and its variations over the very long periods that may have to be considered. Data obtained from short-term corrosion tests, particularly if performed with idealised systems, may therefore be misleading.

To formulate a realistic model of a particular type of deterioration it may be necessary to describe the sequence of processes involved in terms of their rate-determining parameters. Values for these parameters may then be estimated either from related service records or from exposure tests that are designed to simulate the natural environment.

Non-destructive electrochemical methods of corrosion rate measurement can be of great value, when used with appropriate regard to their limitations, as a means of providing data that would otherwise require laborious destructive examinations of large numbers of specimens. Encouraging progress is now being made in developing these techniques for the study of corrosion processes in complex environments.

5 REFERENCES

1. Frohnsdorff, G., L.W. Masters and J.W. Martin. An Approach to Improved Durability Standards for Building Materials and Components. National Bureau of Standards (U.S.) Technical Note 1120 (1980).

2. ASTM E632-82, Standard Practice for Developing Accelerated Tests to Aid Prediction of Service Life of Building Components and Materials. American Society for Testing and Materials (1982).
3. Shreir, L.L. and F.L. LaQue. Corrosion Testing and Determination of Corrosion Rates, in L.L. Shreir, ed., Corrosion, 2 vols. (London, Newnes-Butterworth, 2nd edition, 1976) pp. 20.3-20.105.
4. Pourbaix, M. Atlas d'Équilibres Électrochimiques (Paris, Gauthier-Villars, 1963).
5. Pourbaix, M. Applications of Electrochemistry in Corrosion Science and in Practice. Corrosion Science 14 (1974) 25-83.
6. Hines, J.G. Analysis of Complex Polarisation Curves. British Corrosion Journal 18 (1983) 10-14.
7. Edeleanu, C. and J.G. Hines. Modelling Approach to Corrosion Prediction. British Corrosion Journal 18 (1983) 6-9.
8. Mansfeld, F. New Approaches to Atmospheric Corrosion Research Using Electrochemical Techniques, in R.N. Parkins, ed., Corrosion Processes (London, Applied Science Publishers, 1982) pp.1-76.
9. Tuutti, K. Corrosion of Steel in Concrete. Swedish Cement and Concrete Research Institute (Stockholm) Research Report F04 (1982).
10. Page, C.L. and K.W.J. Treadaway. Aspects of the Electrochemistry of Steel in Concrete. Nature 297 (1982) 109-115.
11. Longuet, P., L. Burglen and A. Zelwer. La Phase Liquide du Ciment Hydraté. Revue des Matériaux de Construction et de Travaux Publics 676 (1973) 35-41.
12. Schiessl, P. Zur Frage der zulässigen Rissbreite und der erforderlichen Betondeckung im Stahlbetonbau - unter besonderer Berücksichtigung der Karbonatisierung des Betons. DAfS, Heft 255 (1976) 39-49.
13. Treadaway, K.W.J., G. Macmillan, P. Hawkins and C. Fontenay. The Influence of Concrete Quality on Carbonation in Middle Eastern Conditions - A Preliminary Study, in A.P. Crane, ed., Corrosion of Reinforcement in Concrete Construction (London, Society of Chemical Industry/Ellis-Horwood, 1983) pp.101-118.
14. Sorensen, B. and E. Maahn. Penetration Rate of Chloride in Marine Concrete Structures. Nordic Concrete Research 1 (1982), 24.1-24.18.
15. Page, C.L., N.R. Short and A. El Tarras. Diffusion of Chloride Ions in Hardened Cement Pastes. Cement and Concrete Research 11 (1981) 395-406.
16. Stern, M. and A.L. Geary. Electrochemical Polarisation I A Theoretical Analysis of the Shape of Polarisation Curves. Journal of the Electrochemical Society 104 (1957) 56-63.
17. Stern, M. and E.D. Weisert. Experimental Observations on the Relation between Polarisation Resistance and Corrosion Rate. Proceedings of the American Society for Testing and Materials 59 (1959) 1280-1291.

18. Andrade, C. and J.A. Gonzalez. Quantitative Measurements of Corrosion Rate of Reinforcing Steels Embedded in Concrete Using Polarisation Resistance Measurements. Werkstoffe und Korrosion 29 (1978) 515-519.
19. Gonzalez, J.A., S. Algaba and C. Andrade. Corrosion of Reinforcing Bars in Carbonated Concrete. British Corrosion Journal 15 (1980) 135-139.
20. Page, C.L. and P. Lambert. Unpublished results.
21. Escalante, E., M. Cohen and A.H. Kahn. Measuring the Corrosion Rate of Reinforcing Steel in Concrete. National Bureau of Standards (U.S.) Report NBSIR 84-2853 (1984).
22. Clear, K.C. Cost Effective Rigid Concrete Construction and Rehabilitation in Adverse Environments. US Federal Highways Administration, FCP Annual Progress Report, Project 4K (1981).
23. Moore, J.F.A. The Performance of Cavity Wall-Ties. Building Research Establishment (U.K.) Information Paper IP 4/81 (1981); see also, Performance of Cavity Wall-Ties, BRE News 54 (1981) 9.
24. BS 1243: 1978 Specification for Metal Ties for Cavity Wall Construction. British Standards Institution. Amendment AMD 3651 (April 1981).
25. Jones, D. and D. Brinn. Zinc Coated Steels in Buildings. (The British Board of Agrément, Garston, Watford, U.K., 1983).
26. Gonzales, J.A. and C. Andrade. Effect of Carbonation, Chlorides and Relative Ambient Humidity on the Corrosion of Galvanised Rebars Embedded in Concrete. British Corrosion Journal 17 (1982) 21-28.

II. APPROACHES TO SERVICE LIFE PREDICTION IN ADVANCED TECHNOLOGIES

APPROACHES TO SERVICE LIFE PREDICTION OF METALS IN THE NUCLEAR INDUSTRY

Gerhart I. Schuëller
Institut für Mechanik
Universität Innsbruck,A-6o2o Innsbruck
Austria

ABSTRACT

A procedure to quantitatively relate target service life of nuclear components, such as pressure vessels, primary piping and containments to reliability is presented. It is shown how this is also related directly to quality control which, consequently affects maintenance procedures, inspection intervals, etc.. Linear damage models as well as fracture mechanical aspects are discussed. Furthermore the effect of corrosion within the framework of the reliability concept is shown. Finally some applications and suggestions for future research are given.

I INTRODUCTION

The problem of service life prediction of components of nuclear power plants is of utmost importance for developing criteria of quality control, maintenance procedures, inspection intervals, etc.. In this context it is referred to large structural components, such as containments, reactor pressure vessels, etc., mechanical components, such as pipes, valves, pumps, etc.. In general terms, age-related effects may in some cases strongly adversly affect the target service life. These effects may be due to wear, corrosion and fatigue or a combination of it. An extensive survey of these age related failure occurrences is given in [1]. These observations reveal a considerable statistical scatter. Therefore only probabilistic procedures provide a sound basis for realistically modeling of the phenomena. For this purpose reliability analysis has been proven to be a most valuable tool [2]. It is of particular advantage that all other possible failure mechanisms, such as ultimate load failure, etc., may be treated within the framework of the reliability concept. Furthermore, by utilizing this concept the significance of the various parameters involved may be determined and then, based on these results, measures with respect to quality control, etc. can be taken, i.e. designed. In reliability theory the problem of service life is generally reformulated, i.e. a finite probability by which a target service life is not met, is to be sought and defined as failure probability which is the complement of the reliability.

In nuclear engineering reliability considerations are utilized with increasing tendency for analysis (see for example [3-5]). These concepts also form the necessary basis for risk studies [6,7].

2 RELIABILITY ANALYSIS

The scope, principles and foundations of reliability theory including numerous applications are discussed in detail elsewhere (e.g. [2,8]). This encompasses modeling procedures of random variables including parameter estimation procedures, etc.. Therefore only some of the principles pertinent to the problem will be reviewed here. In this context it should be pointed out that the foundations of the theory of structural reliability have been developed by *A.M. Freudenthal* [9,10].

Considering that the loading of a structure and its resistance may be described by two variables only, the probability of failure, p_f, under a single load application may be defined as

$$p_f = \int_0^\infty F_R(x) f_S(x) dx \qquad (2.1)$$

where $F_R(x)$ represents the cumulative distribution function of the structural resistance and $f_S(x)$ the probability density function of the load. Type and parameters of the distributions are chosen i.e. derived, based on statistical evidence and physical reasoning respectively. If more than two variables are required to express the randomness of the load and the resistance, the limit state function may be formulated in a more general way, i.e.

$$g(\underline{X}) = g(X_1, X_2, \ldots, X_n) = 0 \tag{2.2}$$

Failure then occurs when the condition $g(\underline{X} \leq 0)$ is met. Various methods to solve equ. (2.1) and (2.2) have been developed so far [11,12]. As structures are designed to sustain the loading for a certain period of time, i.e. the target service life or design life, this measure is not sufficient because it does not reflect the randomness of the load sequence within this time interval, i.e. range. In other words, the probability of survival or reliability is to be defined as a function of time or an equivalent parameter such as the number of load applications.

Measuring the life of a structure in time T the reliability $L_T(t)$ may be defined as

$$L_T(t) = P(T>t) \tag{2.3}$$

where the structural life T is a random variable. The failure probability within a given time interval [0,t] is then

$$F_T(t) = P(T \leq t) = 1-L_T(t) \tag{2.4}$$

From this information various useful parameters can be derived such as the *mortality distribution*, i.e. the probability that the structure fails within the time interval t and t+dt

$$f_T(t)dt = P(t<T\leq t+dt) = \frac{dF_T(t)}{dt} = -\frac{dL_T(t)}{dt} \tag{2.5}$$

and the *hazard function*, which is the conditional probability that the structure fails within this time interval, given that it survived the time t:

$$h_T(t)dt = P(t<T\leq t+dt \mid T>t)$$

$$h_T(t) = \frac{f_T(t)}{L_T(t)} = -\frac{1}{L_T(t)}\,\frac{dL_T(t)}{dt} = -\frac{d}{dt}\ln L_T(t) \tag{2.6}$$

This results in the following relation between the reliability and the hazard function:

$$L_T(t) = \exp[-\int_o^t h_T(x)dx] \quad (2.7)$$

In most cases of practical application - and this holds also for nuclear engineering structures - the number of load occurrences within a given time interval may be described by a stochastic process which may be modeled by either a Gaussian-, Markov- or Poisson type, etc.. Assuming a Poisson process, the probability of the number of load occurrences, n, - for example defined as the malfunction events of a reactor system, or earthquake events, etc. - within a given time interval [O,t] is given by:

$$p_n(t) = p(n|\nu,t) = \frac{(\nu t)^n e^{-\nu t}}{n!}, n = 0,1,2,...\infty \quad (2.8)$$

where ν is the mean rate of the load event occurrence. The reliability function $L_T(t)$ may then be defined by

$$L_T(t) = \int_o^\infty \sum_{n=o}^{\infty} p(n|\nu,t)[F_S(x)]^n f_R(x)dx$$

$$= \int_o^\infty \sum_{n=o}^{\infty} \frac{(\nu t)^n e^{-\nu t}}{n!}[F_S(x)]^n f_R(x)dx \quad (2.9)$$

where $[F_S(x)]^n f_R(x)dx$ is the probability that a structure with resisting strength between x and x+dx will survive n load applications. Summing over all R the structure with a strength distribution $F_R(x)$ will yield the probability by which the structure survives n load applications.

So far the structural resistance has been considered as a time invariant variable. However, physical reality has shown, that, as stated previously, the service life, and consequently the structural resistance may be affected by other factors, for example by fatigue corrosion, etc., which are time dependent processes. Note that this time dependence can also be expressed in number of load applications within a given period of time (s.Fig.1)). This implies that with increasing time, i.e. load applications, the residual structural resistance decreases. This may be modeled by the following relation

$$R(j) = R_o \cdot \psi(j) \quad (2.10)$$

where j defines the respective load events and $\psi(j)$ is

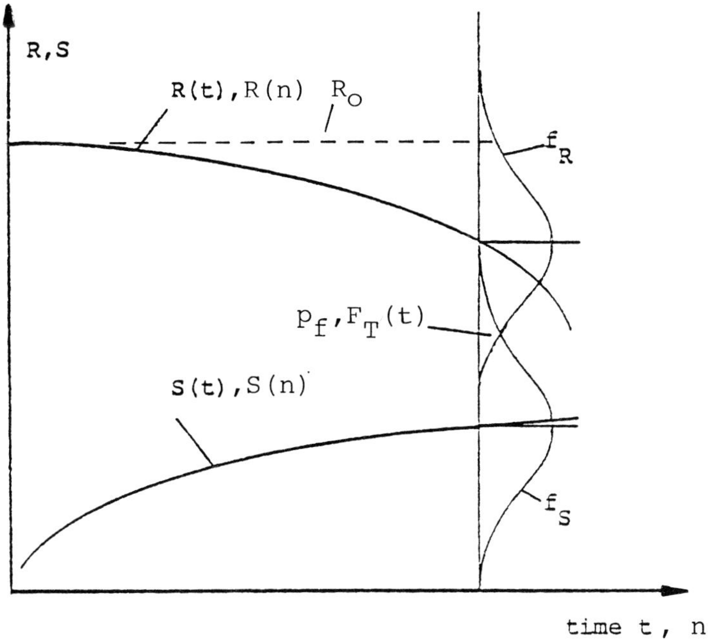

Fig.1: Schematic Sketch of Reliability Concept for Two Random Variables

a decreasing function of the load history. For continuous load processes this is

$$R(t) = R_o \cdot \psi(t) \tag{2.11}$$

$$\psi(t) = \psi[\{S(t)\}] \tag{2.12}$$

A quantitative formulation of $\psi(j)$ i.e. $\psi(t)$ for fatigue and/or corrosion effects may be accomplished by applying for example fatigue or fracture mechanical procedures [13]. Utilizing equ.(2.10)-(2.12) the reliability function can be written as follows

$$L_T(t) = \int_o^{\infty} \sum_{n=o}^{\infty} \frac{(\nu t)^n}{n!} e^{-\nu t} \{\prod_{j=o}^{n} F_{S_j}(x_j)\} f_{R_o}(x_o) dx_o \tag{2.13}$$

where $F_{S_j}(x_j)$ is the distribution function of the load at the j-th load event and $f_{R_o}(x_o)$ the initial structural resistance. It should be pointed out, that the above equation may be also reformulated to take into account initial cracks as observed in almost all welded steel, i.e. metal structures. As there are a larger number of random variables involved in equ.(2.13) it is solved most advantageously by simulation procedures.

Taking into account initial imperfections such as initial crack distribution, equ.(2.13) may be expanded such that the reliability function is

$$L_T(t) = \int_{a_o=o}^{\infty} L_T(t\,|a_o) f(a_o) da_o \tag{2.14}$$

where $f(a_o)$ is the probability density distribution of the initial crack length a_o.

3 DAMAGE MODELS

It has been pointed out that, assuming cyclic loading, the service life of a structure depends on the number of load applications which cause a decrease in structural resistance. This decrease is due to damage effects by each load application.

This effect might certainly be superimposed by an aging effect continuous with time, which, however is for metals of less importance than for plastic type of materials, etc. and hence will not be treated here.

The traditional method to evaluate damage is due to Palmgren [14] and Miner [15] who suggest the hypothesis of a linear stress related damage accumulation, independent of time history, i.e. load sequence. Each load cycle is causing an incremental "damage" D. Failure is defined by $D_f=1$. If the number of load cycles to failure is N_f, the damage caused by each cycle is

$$D = \frac{1}{N_f} \tag{3.1}$$

This type of "damage", of course, lacks a physical relevance and can therefore not be physically interpreted. The cycles to failure, N_f, for each stress range $\Delta\sigma$ is defined by the so called Wöhler curve which is evaluated experimentally. The loss of structural resistance may be calculated by

$$R(n) = R_o[D_f - D(n)] \tag{3.2}$$

Experimental evidence, however, reveals no linear but a progressive decrease of structural resistance. Furthermore, by utilizing this concept, one has to categorize the structural parts under investigation with respect to the quality of initial conditions, such as surface condition, flaws, etc..

The structural resistance of a metal structure will decrease considerably once a crack like flaw, i.e. a fatigue crack has been formed. The amount of the decrease of the structural resistance primarily depends on the size of this crack, the ductility of the material and the geometry. The cause of this decrease of the structural resistance is twofold. On one hand it is due to the trivial decrease of cross sectional area as the crack increases spacially in size and on the other hand due to stress concentration in the vicinity of the crack tip or crack front respectively. Particularly brittle materials are susceptible to failure already at low stress levels as non stable crack growth properties are to be expected.

The application of fracture mechanical concepts enables quantitative description of residual structural resistance taking into account the effects of these parameters on crack growth, i.e. decrease in structural service life. As nuclear structures include weldings, which, despite of the most stringent quality control include initial flaws, only the crack propagation phase on the micro level is therefore of less interest. The crack growth per load cycle may be calculated by [16]

$$\frac{da}{dN} = C(\Delta K)^m \tag{3.3}$$

where

$$\Delta K = \Delta\sigma\sqrt{\pi a}\cdot Y \tag{3.4}$$

is the stress intensity, a the crack length, Y a geometry factor, C and m are material constants. It should be pointed out that equ.(3.3) holds for the linear range only. However, during the service life of a structure the occurrence of extreme loads - which cause plastic deformation - has also to be expected. For this range the crack growth is defined by a strain dependent relation [17]

$$\frac{da}{dN} = A\cdot a\cdot(\Delta\varepsilon_p)^{m_1} \tag{3.5}$$

where A and m_1 are material parameters, $\Delta\varepsilon_p$ the plastic deformation and again, a the crack length. Although there are quite sophisticated methods such as the COD (Crack Opening Displacement) [18] and the J-Integral concept [18] available, the so called Two Criteria Approach [20] or the Feddersen Scheme [21] are used more frequently in practical application to cover the crack growth for the entire load spectrum, i.e. up to the post yield range. The influence of the load sequence for loading conditions of nuclear structures may be neglected [22]. This, of course, does not hold for loading spectra of aircraft structures.

4 CORROSION

Although one of the most sensitive parts of nuclear reactors, the primary piping is for the presently mainly used Pressurized Water Reactors (PWR's) - due to cladding - free of the problem of corrosion, it might effect, however, other components, such as secondary piping systems [23-25], steam generators, reactor coolant pump supports, etc.[26].

It is a well known fact that structural parts may reveal decreasing residual resistance under the influence of corrosive media. This holds for statically as well as dynamically acting loads. These corrosive effects lead to an increase in crackgrowth which in turn causes a decrease in structural resistance and consequently in service life. For static loading the so called *stress corrosion cracking* (SCC) occurs if the stress intensity K, produced by the load, exceeds a certain threshold value (K_{ISCC}). For dynamic loading - depending on the range of stress intensity factor ΔK - the fatigue crack growth may be affected by SCC, or - for low values of ΔK - the crack growth might be increased by vibrational crack corrosion which is also called *fatigue corrosion* (FC). A significant difference between SCC and FC is the fact, that SCC is - given a particular stress intensity present at a crack - a self supporting effect, while FC affects the crack growth only during the time of stress increase within a load cycle. In contrast to SCC, almost no frequency dependence is observed for FC as it is no typical time dependent effect. FC only shows effects when at the crack front a new rupture area develops. By this FC obviously amplifies the mechanical effect of crack growth, i.e. the crack growth taking place in inert environment. For this reason the following relation has been suggested [27] to model crack growth

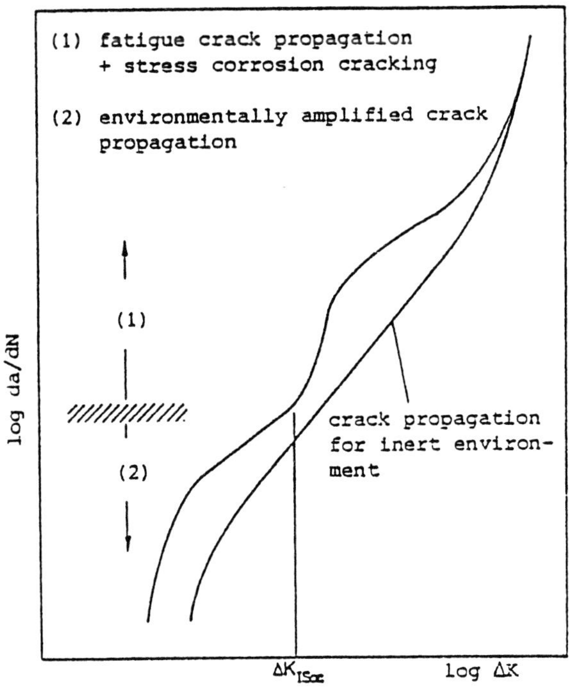

Fig. 2: Influence of Corrosive Media on the Crack Propagation

$$\frac{da}{dN} = C \cdot \Delta K^{3.5} f^{-m} \tag{4.1}$$

It may be observed that the above equation is of simular type as the Paris-Erdogen relation (s.equ.(3.3)). The parameters are valid for SM-50 type steel. The exponent m varies between 0.12 - 0.14 and f is the frequency of the cyclic one step loading. According to [28] the fatigue crack propagation may be superimposed linearly with the effect of SCC. Therefore the following relation holds:

$$\left(\frac{da}{dN}\right)_T = \left(\frac{da}{dN}\right)_{inert} + \int \frac{da}{dt} K(t)dt \tag{4.2}$$

where $(da/dN)_T$ is the total crackgrowth rate, $(da/dN)_{inert}$ the crack growth rate for inert environment, i.e. mechanical fatigue and da/dt the crack growth rate for constant load amplitudes and K(t) the time history of the stress intensity factor.

Referring to equ.(4.1) the corrosion effect is generally taken into account by applying the Paris-Erdogen relation (equ. (3.3)) with an altered parameter C, i.e.

$$\frac{da}{dN} = C_T(\Delta K)^m \qquad (4.3)$$

where $C_T = C \cdot C_{corr}$. This means, that the factor C, which is used for inert environment is multiplied by an additional factor C_{corr}. This factor was shown to be 1.6 or 2.2 for stochastic and constant amplitude loading under salty water conditions.

5 STATISTICAL PROPERTIES OF PARAMETERS

It has already been pointed out previously, that service life of structures in general reveal a significant statistical scatter. This scatter might be explained by analyzing, i.e. investigating the statistical properties of the various input parameters in the reliability and physical models respectively. Uncertainties in service life predictions may arise from several sources: first of all they are due to the variability and uncertainty in external load occurrences, intensity and duration. In case of external hazards, which are important for the design of nuclear power plants, - such as earthquakes - the coefficients of variation of the intensity and duration may be about 1.0. Uncertainties in loadings due to malfunction of plant operations are even larger [24]. Needless to say that additional uncertainties in calculating load effects, i.e. structural modeling uncertainties also contribute to these total load uncertainties. The type of probability distribution and load processes respectively as indicated in equ.(2.9) and (2.13) have to be selected by applying statistical fit procedures and physical reasoning.

The modeling of the structural resistance, i.e. of the material parameters is also quite involved. As already indicated in section 3, the evaluation of damage in quantitative terms includes a number of uncertainties, particularly with respect to the physical interpretation of damage parameters. Although fracture mechanical methods provide a significant improvement in this respect a significant uncertainty still remains. In Table 1 the various parameters involved in the modeling of the structural resistance are listed including those which cause decrease in residual resistance

Parameter	Mean Value	Coefficient of Variation	Type of Distribution
Tensile Stress/StE460	670 N/mm^2	0.04	Weibull
Yield Stress/StE460	520 N/mm^2	0.04	Weibull
Fracture Toughness/StE460	3500 $N/mm^{-3/2}$	0.10	Weibull
Crack Propagation Parameter C			
- for inert Environment	$1.7 \cdot 10^{-13}\ mm^5 \sqrt{mm}/N^3$	0.45	Lognormal
- for corrosive Environment			
for $\Delta K \leq 950 Nmm^{-3/2}$	$1.1 \cdot 10^{-14}$	0.45	Lognormal
for $\Delta K > 950 Nmm^{-3/2}$	$1.9 \cdot 10^{-11}$	0.45	Lognormal
C_{corr} for random load	1.6	-*)	Lognormal
for step loading	2.2	-*)	Lognormal
m	1.5 - 3.7	-	-
Initial Crack Depth a_o	3.65mm	0.93	Special Type

*) no information available

Table 1: Uncertainty Estimates of Material Parameters

along with its stochastic properties. The initial crack distribution plays a very important role with respect to quality control. Utilizing equ.(2.13), (2.14) and (2.4) the target service life is related to F_T, i.e. the probability that this life is not being reached. By additional quality control uncertainties in initial crack lengths may be reduced which causes in turn an increase in reliability estimates. In practical application this procedure is, of course, balanced, i.e. optimized with respect to costs.

6 APPLICATIONS

In nuclear engineering reliability procedures as outlined in the previous sections have already been applied to various components, such as pressure vessels [30], containments [31] and primary piping [24,25]. The problem in nuclear engineering is concerned with meeting, i.e. reaching a specified target service life with a certain probability. This is defined in section 3 as reliability.

In the case of the primary piping it has been shown for example in [24] that, given all possible external and internal hazards, the estimates of the failure probability $F_T(t)$ of the piping system - for a target service life of 40 years - are $5 \cdot 10^{-6}$.

For containments, crack propagation under Loss of Coolant Accident (LOCA) may affect the target service life. But the actual values of failure probabilites are again very small, i.e. in the same order of magnitude as for the primary piping.

In order to reveal the influence of corrosion a structural part of high strength metal, as generally used for nuclear structures, has been analyzed assuming particular external environmental hazard and salty water conditions. The results of the analysis as outlined in section 3 are shown in Fig. 3. It may be observed, that crack growth and corrosion effects increase for example the failure probability within a target service life of 30 years about one order of magnitude.

7 CONCLUSIONS

The analysis as shown in the previous sections shows clearly that recent research in nuclear engineering has

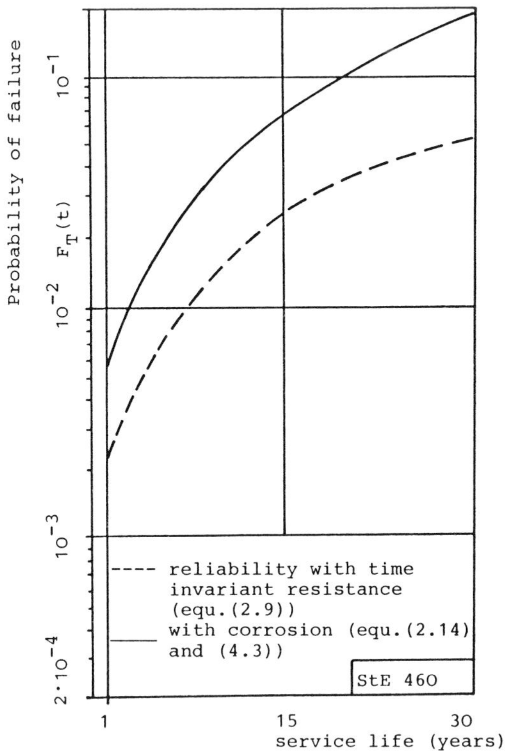

Fig. 3: Effect of Corrosion and Decrease in Residual Resistance (Brittle, High Strength Steel: StE460)

provided a methodology by which the target service life of nuclear structures can be related *quantitatively* to reliability. The probability based concept, i.e. framework, allows the introduction of newly developed physical, i.e. mechanical models without any difficulties. Quality control in terms of initial crack distributions is also being introduced directly in the analysis and - as expected - shows significant effects on the reliability estimates. Less important, but by no means negligible is the effect of corrosion on the service life of nuclear structures. Experimental evidence direct the future attention to small diameter secondary piping. So far the corrosive effect has been considered as an average value. However, more realistic models have to relate the corrosive effects to the respective stress levels occuring during the load history. This problem will be part of further research. Finally, with respect to section 5, it should be pointed out, that additional statistical evidence on the various model input variables is needed in order to reduce the probabilistic model uncertainty as well.

REFERENCES

[1] MURPHY,G.A., GALLAHER, R.B., CASADA, M.L., HOY,H.C.: "Survey of Operating Experience from LERs to Identify Aging Trends - Status Report", NUREG/CR-3543, 1983.

[2] SCHUËLLER, G.I.: "Einführung in die Sicherheit und Zuverlässigkeit von Tragwerken", Wilhelm Ernst und Sohn Verlag, Berlin, 1981.

[3] BENJAMIN, J.R., SCHUËLLER, G.I., WITT, F.J.(Ed.): "Proc. 2nd Int. Seminar on Structural Reliability of Mech. Comp. and Subass. of Nuclear Power Plants", Nucl.Eng.Des., Special Vol., 60, No. 1, Sept. 1980, 171 p.

[4] SCHUËLLER, G.I., BENJAMIN, J.R., COSTES, D., WITT, J.F.(Ed.): "Proc. of 3rd Int.Sem. on Reliability of Nuclear Power Plants", Nucl.Engr. Des., Spec.Vol. 71(3), pp.263-446 (1982).

[5] ANG, A.H.-S., SCHUËLLER, G.I. (Div.Coord.): "Reliability and Risk Analysis of Nuclear Power Plants", Vol.M of SMIRT 7, 1983, North-Holland Publ.Co., Amsterdam.

[6] Reactor Safety Study, WASH1400, U.S.NRC, Washington D.C., Oct. 1975.

[7] Deutsche Risikostudie - Kernkraftwerke, Verlag TÜV Rheinland, Köln, 1979.

[8] ANG, A.H.-S., TANG, W.H.: "Probability Concepts in Engineering, Planning and Design",Vol.II, John Wiley, New York, 1984.

[9] FREUDENTHAL, A.M.: "Safety and the Probability of Structural Failure", Transactions, ASCE, Vol.121, Proc.Paper 2843, 1956, pp.1337-1397.

[10] FREUDENTHAL, A.M., GARRELTS, J.M., SHINOZUKA, M.: "The Analysis of Structural Safety", J.Struct.Div., Proc.ASCE, Vol.92, No.ST1, Febr.1966, pp.267-325.

[11] SHINOZUKA, M.: "Basic Analysis of Structural Safety", ASCE, J.Struct.Eng., Vol.109, No.3, March 1983, pp.721-740.

[12] DOLINSKI, K.: "First-Order Second-Moment Approximation in Reliability of Structural Systems: Critical Review and Alternative Approach", J.Structural Safety, Vol.1, No.3, April 1983, pp.211-231.

[13] OSWALD, G.F., SCHUËLLER, G.I.: "Reliability of Deteriorating Structures", J.Engineering Fracture Mechanics, to appear 1984.
[14] PÄLMGREN, A.: "Die Lebensdauer von Kugellagern", VDI, 68, 1924, pp.339-341.
[15] MINER, M.A.: "Cumulative Damage in Fatigue", J.Appl.Mech., 12, 1945, A159 - A 164.
[16] PARIS, P.C., GOMEZ, M.P., ANDERSON, W.E.: "A Rational Analytic Theory of Fatigue", The Trend in Engineering, 13, 1961, pp. 9-14.
[17] SOLOMON, H.D.: "Low Cycle Fatigue Crack Propagation in 1018 Steel", J. of Materials, JMLSA, Vol. 7, No. 3, Sept. 1972, pp. 299-306.
[18] WELLS, A.A.: "Unstable Crack Propagation in Metals, Cleavage and Fast Fracture", Proc. of the Crack Propagation Symposium Cranfield, 1961, pp.210-230.
[19] RICE, J.R.: "A Path Independent Integral and the Approximate Analysis of Strain Concentrations by Notches and Cracks", J. Appl. Mech. 1968, pp.379-386.
[20] WELLEIN, R.: "Beurteilung der Zuverlässigkeit der Stahlhülle mit Hilfe der Probabilistischen Bruchmechanik", Appendix III, BMFT Report Nor. 150201C, Techn.Univ. München, Sept. 1981, 79p.
[21] FEDDERSEN, C.E.: "Evaluation and Prediction of the Residual Strength of Center Cracked Tension Panels", ASTM STP 468, 1970, pp. 50-78.
[22] OSWALD, G.F.: "On the Reliability of Structures Considering Time Dependent Properties", (in German), Dissertation, Techn. Univ. München, 1983.
[23] PWR Pipe Crack Study Group: "Investigation and Evaluation of Cracking Incidents in Piping in Pressurized Water Reactors", NUREG-0691, Washington D.C., 1980.
[24] SCHÄFER, R., SCHUËLLER, G.I., KAFKA P.: "Probabilistische Untersuchung des Rißfortschrittsverhaltens von Reaktorkomponenten", IABG-Report TF-1605, Ottobrunn, FR-Germany, March, 1984.
[25] HARRIS, D.O. LIM, E.Y, DEDHIA, D.D., WOO H.H., CHOU, C.K.: "Fracture Mechanics Models Developed for Piping Reliability Assessment in Light Water Reactors - Piping Reliability Project", NUREG/CR-2301, VORL-15490, Washington D.C., 1982.
[26] BOHN, M.P., WELLS, J.E., SHIEH, L.C., COVER, L.E., STREIT, R.L.: "An Assessment to Potential Increases in Risk Due to Degradation of Steam Generator Coolant Pump Supports", NUREG/CR-3345, UCID-19719, Washington D.C., 1983.
[27] YOKOBORI, T., SATO, K.: "The Effect of Frequency on Fatigue Crack Propagation Rate and Striation Spacing in 2024-T3 Aluminium Alloy and SM-50 Steel",

Engr. Fract. Mech., 8, 1976.

[28] WEI, R.P., LANDES, J.D.: "Correlation Between Sustained-Load and Fatigue Crack-Growth in High-Strength Steels", Mat. Res. Stand, 9., 1969, p.25.

[29] OBERPARLEITER, W., SCHÜTZ, W.: "Fatigue Life Prediction in a Corrosive Environment", in Integrity of Offshore Structures, Faulkner, D. etal (ed.), The Univ. of Glasgow, Scotland, 1981.

[30] LOMACKY, O., ANG, A.H.-S., AMIN, M.: "Fatigue and Fracture Reliability Analysis of Pressure Vessels", ASME, P.No. 71-PVP-47, Congr. Press. Vessel and Piping, May, 1971.

[31] SCHUËLLER, G.I., KAFKA, P., WELLEIN, R, WITTMANN, F.H., ZORN, N.F., BERNING, A., BOULAHDOUR, T., MIHASHI, M., BAUER, J.: "Zuverlässigkeitsbeurteilung für den Sicherheitseinschluß am Beispiel des Druckwasserreaktors", BMFT-Report 150 201-C, München, Sept. 1981.

A DAMAGE TOLERANCE APPROACH FOR STRUCTURAL INTEGRITY OF AEROSPACE VEHICLES

Frank D. Adams and Robert M. Engle Jr

Wright Aeronautical Laboratories
Wright Patterson AFB, Ohio, USA

ABSTRACT

In the late 1960's the United States Air Force shifted its philosophy of design for aircraft structures from a safe-life approach based on fatigue analysis and scatter factors to a damage tolerance approach (assume existence of initial flaws) based on fracture mechanics technology with emphasis on structural life management through individual aircraft tracking. An entire set of requirements and procedures has been assembled under the Aircraft Structural Integrity Program. This program encompasses analyses, tests, design and inspection procedures to ensure that premature failures will not occur during the design service life of the aircraft. This paper includes discussions on the procedures developed to inspect for structural damage, track the accumulation of damage, and manage the usage of the aircraft to minimize the rate of damage accumulations for the entire force of aircraft. This force management program is described in detail with examples at all levels from basic design through fleet management.

INTRODUCTION

The purpose of this paper is to outline and summarize the current approach used by the United States Air Force (USAF) to design damage tolerant aircraft structure. This material was extracted from the referenced papers and reports, and organized for presentation to the NATO Advanced Research Workshop on "Problems in Service Life Predictions of Building and Construction Materials." Although the approach discussed here has been specialized to aircraft structure to meet particular requirements of the USAF, the overall philosophy of designing damage tolerant structure has potential application in many industries, especially where structural failure results in a safety problem. It is the authors' desire to convey the following material as a particular application of a more general approach to structural design. To fully appreciate the concept and to set a frame of reference for the reader, a bit of history is presented.

The Air Force has always sought to increase the performance of the aircraft in its fleet. As a result, designers have continually been faced with formidable problems. Greater engine power, improved aerodynamics, and lighter structures all provided increased performance but also required improved structural efficiency. The quest for lighter, more efficient structures has led to the development of alloys with higher yield strengths to permit designers to reduce section properties and, thereby, overall structural weight. The Air Force recognized that aircraft designed to these stringent requirements would have greatly reduced safety margins. Hence, the USAF established the Aircraft Structural Integrity Program (ASIP) which was formalized (1) in the mid 1960's. The emphasis turned from designs based on initial static strength to establishing structural fatigue lives and monitoring the structure throughout its lifetime.

During the 1960's, the Air Force used a "safe life" approach for structural design to ensure safety and durability. In general, the fatigue life of the structure was determined through testing of materials, components and full scale airframes and a safety factor of 4 was applied to reduce the probability of failure to an acceptable value. The full scale fatigue test was performed using a production aircraft. The results of this test were used to establish aircraft inspection times and schedules for modifications and repairs. This approach did not explicitly take into account the possibility of infrequently occurring flaws due to material defects and manufacturing processes.

In December 1969 a catastrophic crash of an F-111 aircraft which had only 100 hours of flight service occurred. The cause of the accident was an undetected flaw in the wing pivot fitting which resulted in the loss of the wing during flight. The F-111 crash stimulated the Air Force into adopting a new design philosophy. The basic assumptions in this new approach are: flaws are present in new structures and structures can be designed such that cracks can grow from these flaws for a specified period of unrepaired service without structural failure occurring. This concept is generally referred to as damage tolerant design.

THE CURRENT AIR FORCE APPROACH

The current approach is based on the technology of fracture mechanics. The aircraft structure is no longer assumed to be flaw free; initial cracks are postulated at defined locations. Durability and damage tolerance are defined in terms of flaw sizes and crack growth intervals. Durability is defined as the ability of the aircraft to resist cracking, corrosion, wear and other degradation for the entire design service life (2). Damage tolerance is defined as the ability of the aircraft to resist failure due to the presence of flaws, cracks, or other damage for a specified period of time (2).

A very significant feature of the Air Force's approach, which allows it to be applied to structural design in an economic way, is that structural safety (avoid catastrophic failure in flight) and durability (avoid expensive modifications and retrofits) are decoupled. Each is considered separately and has its own set of requirements. This was not the case with the "safe life" approach. This approach assumed that the aircraft structure was free of flaws and defects and all life analyses were performed using fatigue approaches with scatter factors to account for material variations, initial quality variations, and environmental effects. Nothing was included in the analyses or design requirements to provide protection from the effects of pre-existing or service-induced flaws. Further, no differentiation was made, in either analysis or test, as to the type of damage causing the failures. Thus, neither fracture-critical structure nor durability-critical structure were identified.

The current Aircraft Structural Integrity Program is firmly embedded in Air Force regulations. USAF Regulation 80-13 requires the System Program Offices (they have responsibility for developing new aircraft weapon systems) to comply with MIL-STD 1530A (2). This document defines the overall requirements necessary to achieve structural integrity of USAF airplanes, and specifies acceptance methods of contractor compliance. The standard is used by contractors in conducting the development of an airframe for a particular weapon or support system, and by government personnel in managing the development, production, and operational support of a particular airplane system throughout its life cycle. Required tasks are outlined in Figure 1 and are discussed later in the paper. MIL-STD-1530A is directly applicable to manned power-driven aircraft having fixed or adjustable fixed wings and to those portions of manned helicopters which have similar structural characteristics. Helicopter transmission systems and rotors, and other dynamic machinery including engines are not covered by this standard. Detailed requirements related to damage tolerance, durability, structural testing, sonic fatigue, etc. are provided in a set of military specifications.

TASK I	TASK II	TASK III	TASK IV	TASK V
DESIGN INFORMATION	DESIGN ANALYSES AND DEVELOPMENT TESTS	FULL SCALE TESTING	FORCE MANAGEMENT DATA PACKAGE	FORCE MANAGEMENT
ASIP MASTER PLAN STRUCTURAL DESIGN CRITERIA DAMAGE TOLERANCE & DURABILITY CONTROL PLANS SELECTION OF MAT'LS, PROCESSES, & JOINING METHODS DESIGN SERVICE LIFE AND DESIGN USAGE	MATERIALS AND JOINT ALLOWABLES LOAD ANALYSIS DESIGN SERVICE LOADS SPECTRA DESIGN CHEMICAL/ THERMAL ENVIRONMENT SPECTRA STRESS ANALYSIS DAMAGE TOLERANCE ANALYSIS DURABILITY ANALYSIS SONIC ANALYSIS VIBRATION ANALYSIS FLUTTER ANALYSIS NUCLEAR WEAPONS EFFECTS ANALYSIS NON-NUCLEAR WEAPONS EFFECTS ANALYSIS DESIGN DEVELOPMENT TESTS	STATIC TESTS DURABILITY TESTS DAMAGE TOLERANCE TESTS FLIGHT & GROUND OPERATIONS TESTS SONIC TESTS FLIGHT VIBRATION TESTS FLUTTER TESTS INTERPRETATION & EVALUATION OF TEST RESULTS	FINAL ANALYSES STRENGTH SUMMARY FORCE STRUCTURAL MAINTENANCE PLAN LOADS / ENVIRONMENT SPECTRA SURVEY INDIVIDUAL AIRPLANE TRACKING PROGRAM	LOADS / ENVIRONMENT SPECTRA SURVEY INDIVIDUAL AIRPLANE TRACKING DATA INDIVIDUAL AIRPLANE MAINTENANCE TIMES STRUCTURAL MAINTENANCE RECORDS

Figure 1. USAF Aircraft Structural Integrity Program Tasks (Ref. 2).

The damage tolerance requirements, detailed in MIL-A-83444, basically consist of: identification of fracture critical parts; initial flaw size assumptions; minimum residual strength requirements; minimum periods of unrepaired service usage; and damage growth limits in terms of aircraft usage (3). The damage tolerance requirements and the methodology needed to apply them are described in detail in the USAF Damage Tolerant Design Handbook (4). A significant element of MIL-A-83444 is the fastener policy which allows the designer to take advantage of certain gains from modern fasteners but provides a factor of safety in the event the fasteners are not completely effective.

The durability requirements, established in MIL-A-8866B, are couched in terms of economic life (5). MIL-A-8866B requires that the economic life be greater than the design service life. In general, the economic life is defined as the occurrence of widespread damage which is uneconomical to repair and could lead to functional impairment if not repaired. The currently available approaches to satisfying the durability requirements are contained in the USAF Durability Handbook (6). Conceptually, durability analysis differs from damage tolerance analysis in that it deals with a large number of structural elements and the overall distribution of damage in the entire structure rather than with a single flaw at a specified location. This suggests a probabilistic approach as opposed to a deterministic approach.

THE AIRCRAFT STRUCTURAL INTEGRITY PROGRAM

ASIP defines a philosphy for protecting and managing the structural integrity of the USAF fleet. Once the assumption of a flaw free airplane is set aside, the tasks outlined in Figure 1 become a logical progression. In the next few paragraphs each of the five tasks will be discussed and some examples will be given. The damage tolerance design philosophy has not been restricted to aircraft designed since 1974. Starting with the F-111 recovery program (7), the Air Force has performed durability and damage tolerance assessments (DADTA) of virtually every aircraft type in the inventory. Each DADTA applied every step in Figure 1 to whatever level was considered appropriate for the individual aircraft being evaluated.

Task I: Design Information. The objective of Task I is to develop criteria which must be applied during the design so that the specific requirements are met. Of first order is the development by the contractor of an ASIP Master Plan for any new aircraft system. This plan includes: time-phased scheduling; task integration; unique features of the aircraft system; anticipated problems; and impact of delays and recovery plans. It is updated annually or whenever important changes occur and must be approved by the Air Force. The contractor must also establish structural design criteria. These must be in concert with the applicable military specifications discussed previously.

Development of the durability and damage tolerance control plans is a vital part of Task I. In terms of actual application, these two plans are often combined into a single plan. The function of this plan is two-fold; identify the critical parts and define control of these parts. Landy and Smithers (8) presented a generalized model (Figure 2) of a durability and damage control plan. Critical parts are chosen according to established criteria after a design is finalized. Additional parts are usually added later based on non-engineering criteria such as high cost to repair or mission impairment. Once parts have been identified as critical, they become the responsibility of the Durability and Damage Tolerance Control Board. This is an interdisciplinary board (engineering, quality control, manufacturing, NDI, etc.) which administers the specifics of the durability and damage control plan. For instance, the F-16 program (9) designated forty-eight parts as fracture-critical and wrote a separate specification detailing requirements, traceability, manufacturing and processing for each of these parts.

Task II: Design Analyses and Development Tests. In Task II, material and joint allowables are developed. Load analyses are performed to obtain the design load spectra (load histories). Thermal and chemical environments are considered. This task also includes the stress analysis work. In addition, fracture mechanics life analyses are required for all structure on the fracture critical parts list. Life analyses are performed on durability critical parts. Other analyses include sonic fatigue, vibration, flutter, nuclear and non-nuclear weapons effects. Developmental testing to verify all of these analyses is performed under Task II.

The magnitude of the work required by Task II seems overwhelming at first glance. However, procedures and policies (10) have been developed to render the task more manageable. Buntin (9) outlines the procedure used for the F-16, which is illustrated in Figure 3. First the random flight-by-flight loads spectrum is developed from mission analysis parameters. This is then converted to a stress history for each fracture critical location (control point). Concurrently, basic material properties are developed. These data are used to verify the basic crack growth model and the stress intensity factor solutions. Once the overall model correlation has been performed, all the elements are combined to obtain the life predictions and inspection interval calculations for each control point.

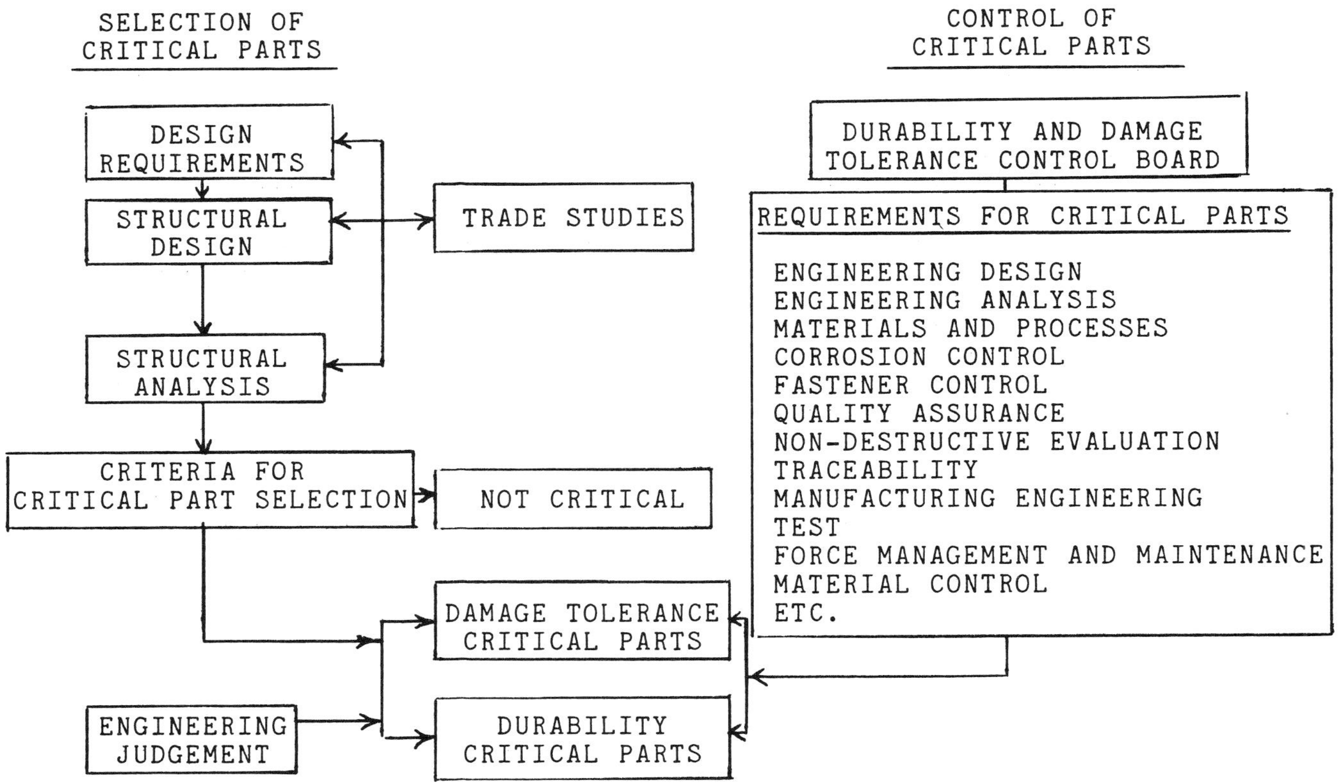

Figure 2. Generalized Durability and Damage Tolerance Control Plan (Ref.9)

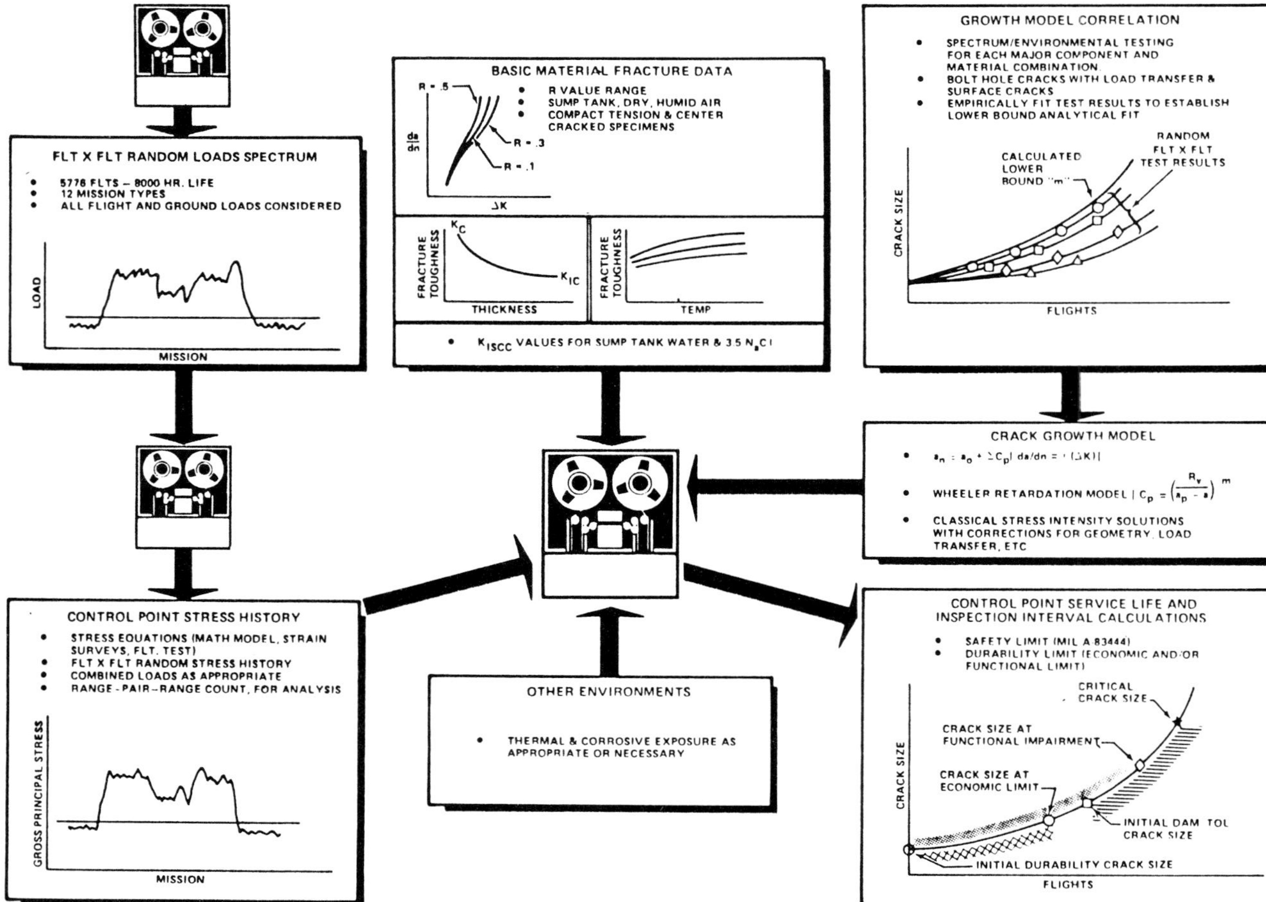

Figure 3. F-16 Safety and Durability Design Methodology (Ref. 9).

Design development testing is conducted at the coupon and structural element level. Purposes of this testing include: 1) analysis verification for both residual strength and crack growth; 2) evaluation of test spectrum truncation levels and procedures; and 3) detailed simulations of isolated design features. An example of the extent to which design development testing may be utilized was demonstrated by Swift and Wang (11) in a program looking at fuselage structure. Fifteen stiffened panels ranging in size from five feet by ten feet to six feet by fourteen feet were tested to evaluate effects on structural strength and life of skin materials, stiffener spacing and design, crack stoppers and curvature. Analysis models for both residual strength and crack growth were verified using these data. The Boeing Commercial Airplane Company has formalized their design development testing data base into a living corporate memory called the Boeing Durability System (12).

Task III: Full Scale Testing. This task includes static strength tests of the full airframe and/or selected components. Cyclic durability (fatigue) tests are also performed. Damage tolerance testing is accomplished on structures in which flaws have been introduced at critical locations. The basic objectives of all full scale testing is to verify analyses performed during Task II. This is markedly different from the "safe life" approach used prior to 1970 which used full scale testing to determine the structural life.

Full scale testing, especially fatigue testing, represents some of the more significant differences in the current design philosophy relative to the old "safe life" procedures. Coffin and Tiffany (13) presented an excellent overview of the test policies and procedue. To highlight some of the more significant features of the testing requirements consider the following:

1) The full scale test aricle will be an early Research, Development, Test, and Evaluation (RDT&E) aircraft, not a production aircraft. However, it should be as representative as possible of the operational airframe.

2) A flight-by-flight loads spectrum typical of projected operational usage will be used, as opposed to the block loading used previously.

3) One lifetime of fatigue testing will be completed prior to production go-ahead and two lifetimes prior to delivery of the first production aircraft. The old policy did not specify scheduling of the fatigue test; consequently results often were obtained too late to prevent costly redesign and retrofits.

4) A teardown inspection will be performed on the full scale test article. Fractographic evaluations of the fatigue cracks may be performed to provide additional data for correlation with strain gages and analyses.

Many options are available in the overall test requirements to tailor the tests to the individual weapon system. In addition to the RDT&E airframe, provisions exist for testing structural elements to isolate effects of specialized structures. The emphasis of the full scale testing task is much more oriented toward verification than it was previously. Data from the full scale test program are used to identify "hot spots," verify analyses, demonstrate compliance with the damage tolerance requirements, assess the adequacy of the inspection procedures and determine the initial quality of the structure.

Task IV: Force Management Data Package. Tasks IV and V involve Force Management which is the term the USAF employs in dealing with the assurance of structural integrity after the aircraft is operational. This assurance is maintained by monitoring and tracking the usage of each individual aircraft. Task IV is performed by the contractor while Task V is accomplished by the Air Force.

In Task IV the contractor is required to develop a Force Structural Maintenance Plan (FSMP). This is the "owner's manual" and provides plans for the WHAT, WHEN, HOW and WHERE actions needed during the service life. Cost factors are also developed during this task. The plans and data package required to perform a "Load/Environment Spectra Survey" (L/ESS) are also established. This involves instrumenting a portion of the aircraft fleet with multi-channel recorders to measure actual flight parameters to update design load histories. These data are later used to identify changes in operational load histories and to update life analyses performed using the design load histories. A significant part of Task IV is the development of an individual airplane tracking (IAT) program. Most of the current operational aircraft in the USAF fleet are tracked to determine individual differences in usage which may affect structural life. Simple instrumentation such as counting accelerometers or pilot/engineer log data are used for this purpose. The USAF Handbook of Force Management Methods (14) presents the entire force management process in detail.

As with the durability and damage tolerance control plans, the L/ESS and IAT program plans are combined into the Force Structural Maintenance Plan to provide a single controlling document. The FSMP is a constantly changing document which is kept current through feedback from the IAT program. Content of the FSMP will vary between aircraft types but as a minimum should contain (14):

1) Specification of all anticipated inspection, repair, and modification activity in a time phased format with provision for individual aircraft scheduling.

2) Summary of critical locations and values of accumulated damage which will trigger required maintenance actions.

3) Supporting data required for implementation of the maintenance procedures.

Figure 4 demonstrates the manner in which usage data from an individual aircraft are combined with data from the FSMP to determine the amount of the design service life expended. The coordinated tracking of damage accumulation in each individual aircraft and the associated scheduling comprise the Force Management task performed by the Air Force.

Task V: Force Management. Task V is the Force Management program accomplished by the Air Force. This involves implementation of the Force Structural Maintenance Plan. Information from the Loads/Environment Spectra Survey is used to update this plan. Data from the Individual Airplane Tracking program are employed during the life of the aircraft to account for individual usage. This task also involves the development and control of all maintenance records.

The implementation of the Force Management task requires coordination between several organizations. The Aeronautical Systems Division of the Air Force Systems Command (AFSC/ASD) is responsible for management of the entire ASIP function until the aircraft system is transferred to the operational command. At this time, responsibility falls to the Air Logistics Centers (ALC) of the Air Force Logistics Command. There are five ALCs, each of which has responsibility for a certain number of aircraft systems. They implement both the L/ESS and IAT programs and manage both the data acquisition and transmittal to a central repository. This central data bank is called the Aircraft Structural Integrity Management Information System (ASIMIS) and resides at Oklahoma City Air Logistics Center. Data sent there are reduced, collated and assembled into force management reports for each aircraft system in the inventory. These reports are then sent to the individual system managers at the appropriate ALC for use in scheduling maintenance and/or modification actions for individual aircraft. Figure 5 illustrates the ASIP function breakdown into the separate elements of the Force Management task and outlines the responsibilities for each action. AFSC/ASD and the airframe

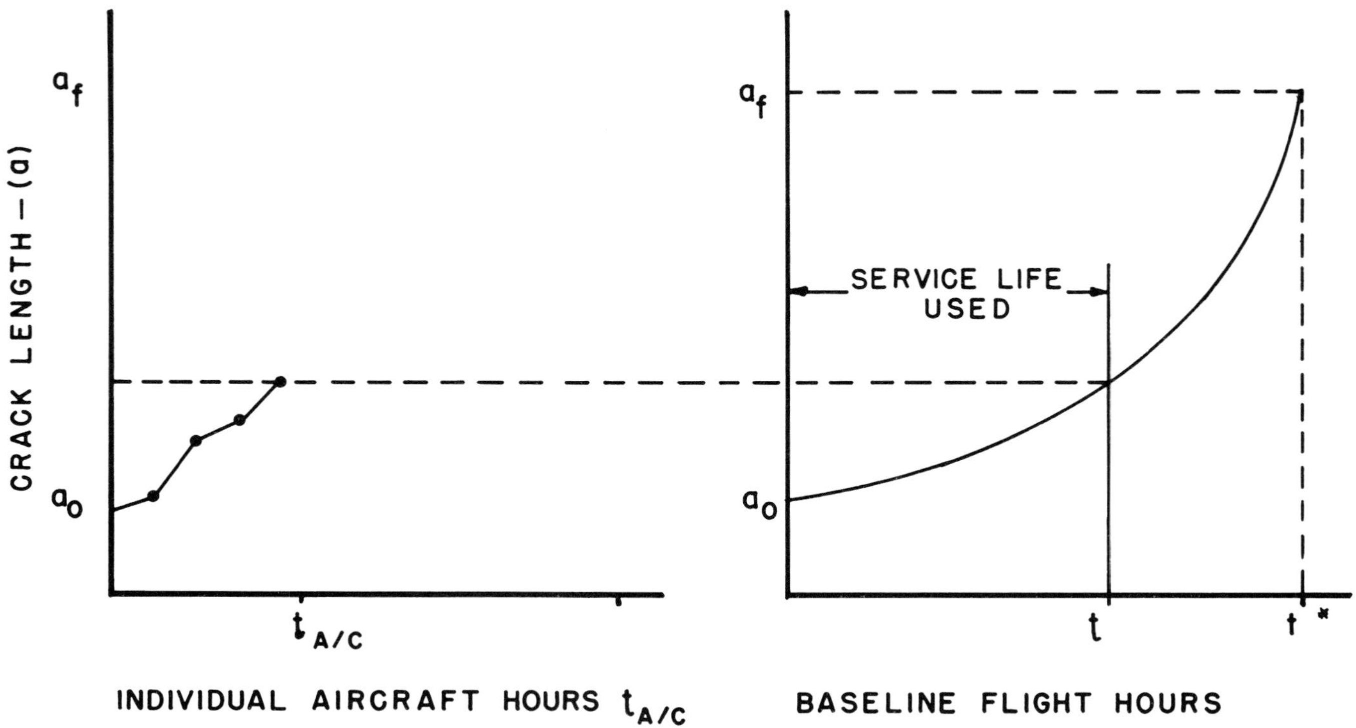

Figure 4. Relating IAT Usage to FSMP Usage (Ref. 14).

Force Management Element	Management Responsibility	Work Responsibility
1. Final Analysis	AFSC/ASD	Contractor
2. Strength Summary	AFSC/ASD	Contractor
3. Force Structural Maintenance Plan	AFSC/ASD	Contractor
4. L/ESS Data package (Task IV)	AFSC/ASD (ASIMIS)	Contractor
5. IAT Data Package (Task IV)	AFSC/ASD	Contractor
6. L/ESS (TASK V)	AFSC/ASD or AFLC/ALC	
a) Data Acquisiton Procedures		AFSC/ASD
b) Data Acquisition Services		Operating Command
c) Training		AFSC/ASD, AFLC/ALC and Operating Commands
d) Data Processing		AFLC/ASIMIS
e) Data Evaluation		AFSC/ASD or AFLC/ALC
7. IAT (Task V)	AFSC/ASD or AFLC/ALC	
a) Data Acquisition Procedures		AFSC/ASC
b) Data Acquisition Services		Operating Command
c) Training		AFSC/ASD,AFLC/ALC and Operating Commands
d) Data Processing		AFLC/ASIMIS
e) Data Evaluation		AFSC/ASD or AFLC/ALC
8. Individual Aircraft Maintenance Times	AFLC/ALC	AFLC/ALC
9. Structural Maintenance Records	AFLC/ALC	AFLC/ALC and Operating Commands

Figure 5. Organizational Responsibilities for Force Management Elements. (Ref. 14)

contractors maintain a close relationship with the ALCs throughout the life of the aircraft system to provide any needed consultation or support.

The heart of Force Management is the IAT program, whose purpose is to monitor the usage of the aircraft system throughout its service life. The IAT program provides periodic estimates of potential crack lengths at critical locations for each airplane; establishes a crack growth-life curve for the anticipated usage; estimates future damage accumulation rates and flying severity; and monitors loads parameters to provide information on usage variations from base to base and aircraft to aircraft. These data provide the system manager with the information necessary to schedule maintenance actions and rotate aircraft to different bases to normalize damage accumulation rates.

The L/ESS function gathers loads data from operational aircraft to update the design service spectra to reflect actual usage. The recorded loads are compared to the design loads. If the variations become significant, the durability and damage tolerance analyses may have to be updated. Possibly the FSMP may require revision. Updating DADTA and the FSMP are not tasks required under L/ESS, but they may be initiated based on L/ESS data. A secondary function of the L/ESS program is to provide data for the generation of structural design criteria. The desired usage of a new aircraft system is defined in terms of mission profiles. Design criteria provide load frequency and magnitude information for various mission profiles for each different type of aircraft. These data are made available as number of exceedances of a given load level, characterized by n_z, the normal load factor, for a given time, usually per one thousand flight hours. Figure 6 is a typical exceedance curve for a supersonic trainer. Combining this n_z distribution with a mission profile and a desired service life represents the starting point for developing a design service load history. The typical exceedance curves for various classes of aircraft are presented in reference 5.

SUMMARY

The Aircraft Structural Integrity Program addresses the "cradle to grave" requirements for the design, analysis, test, and force management of USAF aircraft. Structural safety is provided through damage tolerant design and test verification. The full scale testing policy permits identification and correction of durability problems prior to full production go-ahead. Inspection and modification requirements are defined through the Force Structural Maintenance Plan based on the design usage spectrum. This plan is updated based on actual aircraft usage measured by the

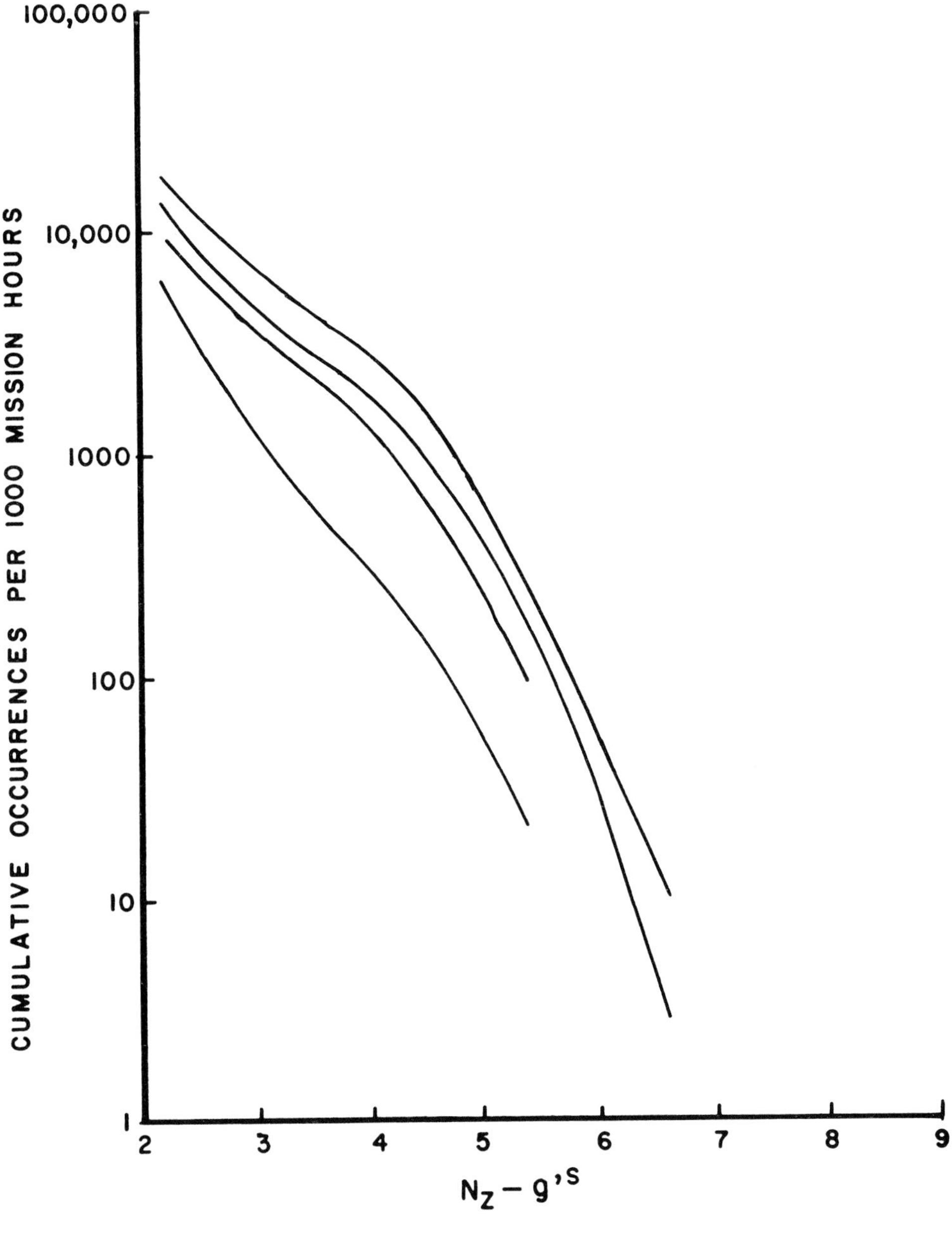

Figure 6. Sample Normal Load Factor Spectra (Ref. 14).

Loads/Environment Spectra Survey. Maintenance times are adjusted for individual aircraft using data from the Individual Aircraft Tracking Program.

This design philosophy is being applied to all new aircraft in the USAF inventory through the updated system of design requirements. Durability and damage tolerance assessments have also been conducted on virtually all the older aircraft types in the USAF fleet. The Damage tolerance specifications for composite materials are currently under development. At present design procedures for composite structures are based on proof tests but still adhere to the principles which underlie MIL-STD-1530A. In additions the principles of damage tolerant design and structural life management are being extended to include secondary structure, helicopters and even aircraft engines.

References

1. Wells, H.M., "Air Force Structural Integrity Program Requirements," ASD-TR-66-57, Jan 1968.

2. Department of The Air Force, "Aircraft Structural Integrity Program, Airplane Requirements," MIL-STD-1530A, Dec 1975.

3. Department of the Air Force, "Airplane Damage Tolerance Requirements," MIL-A-83444, Jul 1974.

4. Gallagher, J.P., Gieseler, F.J., Berens, A.P. and Engle, R.M., "USAF Damage Tolerant Design Handbook: Guidelines for the Analysis and Design of Damage Tolerant Aircraft Structures," AFWAL-TR-82-3073, Jan 1984.

5. Department of the Air Force, "Airplane Durability Design Requirements," MIL-A-8866B, Aug 1975.

6. Manning, S.D. and Yang, J.N., "USAF Durability Design Handbook: Guidelines for the Analysis and Design of Durable Aircraft Structures," AFWAL-TR-83-3027, Jan 1984.

7. Buntin, W.D., "Concept and Conduct of Proof Test of F-111 Production Aircraft," presented to the Royal Aeronautical Society, London, United Kingdom, Oct 1971.

8. Landy, M.A. and Smithers, O.L., "Durability and Damage Tolerance Control Plans for USAF Aircraft," Journal of Aircraft, Vol. 20, No. 8, Aug 1983, pp 689-698.

9. Buntin, W.D., "Durability and Safety in the F-16 Airframe," Structural Integrity Technology, ASME, May 1979.

10. Wood, H.A., "Fracture Control Procedures for Aircraft Structural Integrity," AFFDL-TR-71-89, Jul 1971.

11. Swift, T. and Wang, D.Y., "Damage Tolerant Design-Analysis Methods and Test Verification of Fuselage Structure," Air Force Conference on Fatigue and Fracture, AFFDL-TR-70-144, Sept 1970.

12. Goranson, U.G., Hall, J., Maclin, J.R. and Watanabe, R.T., "Long Life Damage Tolerant Jet Transport Structures," Design of Fatigue and Fracture and Fracture Resistant Structures, ASTM STP 761, 1981.

13. Coffin, M.D. and Tiffany, C.F., "New Air Force Requirements for Structural Safety, Durability, and Life Management," Journal of Aircraft, Vol. 13, No. 2, Feb 1976, pp 93-98.

14. Berens, A.P., Morcock, D.S., Brown, K.E., Davis, J.C., Johnson, R.L., Larson, C.E., and White, D.J., "Handbook of Force Management Methods," AFWAL-TR-81-3079, April 1981.

III. COMMONALITIES BETWEEN SERVICE LIFE PREDICTION PROBLEMS IN BUILDING/CONSTRUCTION TECHNOLOGY AND ADVANCED TECHNOLOGIES

ESSENTIAL DATA FOR SERVICE LIFE PREDICTION

Göran Fagerlund

CEMENTA AB, P.O. Box 144, S-182 21 DANDERYD, Sweden

ABSTRACT

Durability design based upon old experience-codes of good practice etc - is still widely used. It is, however, a dangerous method that has led to a lot of durability problems; examples are given in the report. A design based upon a service life calculation has many advantages but is very complex. A lot of data concerning functional requirements, environment, deterioration mechanism and material are needed. Examples of relatively elaborate methods for the estimation of the service life of concrete are presented briefly. The use of the service life concept means that the traditional accelerated test methods must often be abandoned in favour of non-accelerated tests.

1 TODAY'S SITUATION

When a building designer selects the materials that shall be included in a certain building component - e.g. a floor, an external wall, a roof etc - and when he selects the way to combine the different materials to form the finished component, it is fairly easy for him to satisfy the required function of the component at "time zero"; i.e. at the time of delivery of the finished building. He can base his choice of materials and his design on reasonably welldefined properties that are listed in product brochures from the producer or utilize general data listed in handbooks etc; such properties include the 28-day strength of the concrete or the heat transfer coefficient of the heat insulation. The considerably more complex properties that describe the "ageing" of the material in situ, i.e. the time-dependent interaction between two adjacent materials, or the interaction between the material and the surrounding environment, or the effect of possible changes in the structure or in the environment, are of no importance for the function at time zero. Those properties are also very seldom accounted for by the producer or in handbooks. In many cases, the producer and the designer do not even know which material properties are of importance for the long-term function.

Carpets to be used on concrete floors on the ground can be taken as trivial examples of products for which important information is still lacking to a great extent. Properties such as colour, thickness etc are accounted for by the manufacturers but properties that are extremely important for the long term function are normally unknown; e.g. the (possibly time -dependent) relation between the relative humidity (RH) and the moisture diffusivity of the carpet, or the interaction between the carpet and different glues and putties in the presence of alkaline moisture from the concrete. The effect of the permeability of the carpet on the moisture state of the glue and putty is shown in Fig. 1. The chart is valid for one-family houses with a concrete slab directly on the ground, /1/. The RH at the interface will be 90% if the carpet has a diffusivity at its present moisture state that is 1/50 of that of the concrete slab but only 82% if the diffusivity ratio is 1/5 (assuming a heat insulation of the floor that gives a temperature drop of 1.5 ^{o}C and an RH value in the room of 40%). This difference in RH is dramatic; most putties and glues are durable at RH = 82% but very few at RH = 90% (in fact, however, the effect of the RH on the durability of glues and putties is also unkown to a large extent. There are certainly materials (organic) that do not even endure an RH as low as 80% for a very long time as has been experienced in Sweden).

Another example of missing information, valid for the same type of structure, deals with the moisture absorption in the heat insulation that should normally be laid beneath the concrete in order to lower the RH value of the flooring material; Fig. 1. It is very important that this layer does not absorb water. If it does, the RH value of the concrete could reach too high a value. A material often used in Sweden is expanded clay granules. A layer of about 25 cm should theoretically lower the RH value in the floor to about 80% providing there is no capillarity in the layer. Previously, there were very few problems in Sweden with such layers. Nor did simple capillarity tests with layers of expanded clay give reason to believe that the material absorbed water to a height of more than a few centimetres. The manufacturer therefore stated that his material had "no capillary suction", and this was believed by the designers. Then, over a period of some years, a large amount of moisture-induced damage occurred in floors with that type of insulation and covered with plastic carpets. All measurements indicated that moisture must have been transferred from the ground to the concrete slab via the insulation. A theoretical analysis /2/ showed that even an extremely slow capillary suction, for example in dust grains on the surface of the granules, was enough to raise the RH value of the floor to an unacceptable level for the flooring material; the diffusion downwards caused by the thermal gradient and the drying upwards through the very dense carpet was not enough to compensate for even a slow capillarity. This slow capillary suction could not be detected by the rather coarse method used for testing the capillarity. Consequently the producer could be said to have been in "good faith". Quite a new measurement principle was suggested by which even a very slow capillary suction could be observed /2/. Unfortunately, this method has never come into use and the manufacturer still claims that his material has "no suction". Fortunately, however, the material is very little used due to the poor experience obtained, but obtaining that experience, which could have been predicted if the material had been correctly tested, has cost a lot of money and caused much suffering among residents.

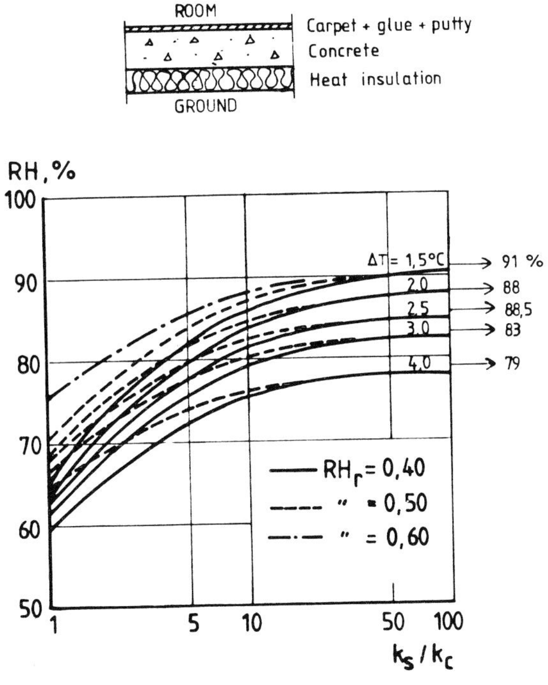

Fig. 1 The equilibrium relative humidity at the interface between a carpet and a concrete slab on the ground /1/. (ΔT = the temperature drop across the floor, RH_r = RH in the room above the floor, k_s and k_c are the moisture diffusivities of the concrete slab and the carpet respectively).

There are of course many more examples to give of durability problems that have been caused by ignorance of data, lack of data or even misleading data.

This evident lack of interest in the long term performance of a building during its design phase is rather remarkable. It reflects the considerable complexity involved in dealing with durability and service life predictions, especially the very great difficulties in making accurate service life calculations. Instead of actually designing the building with regard to its durability one simply turns to experience of the behaviour of similar buildings in similar environments. This experience is normally expressed in terms of simple rules in different official standards or in codes of good practice. Such rules are certainly quantitative but do not deal with service life directly but merely with maximum or minimum values of certain substituting properties that are supposed to be coupled to service life.

Concrete can be used as an example of a material for which substituting properties are widely used. The durability requirements are normally expressed in terms of requirements for:

- compressive strength
- water-cement ratio
- permeability to water (sometimes to air)
- air-content of the fresh concrete
- concrete cover over the reinforcement
- cement type

Many of those properties are measured on specially manufactured specimens and not on the concrete in the erected structure. This is also typical of materials other than concrete. In many cases, the specimens used for quality control are not even very representative of the structure. The consequences of a design based upon experience only will be discussed below after which the possibilities of an alternative design based upon the service life concept will be presented.

2 EXPERIENCE AS A BASIS FOR SERVICE LIFE DESIGN

A design based solely upon experience of the behaviour of similar structures has many drawbacks. It is very defensive and dangerous in a situation where the design principles are often changed more or less radically and where more and more new building materials and components are marketed and utilized rather uncritically.

In Sweden, there are many examples of changes in building construction that have caused great problems during the last decades, simply because no-one ever analyzed the consequences of such changes on the service life. Therefore the old design rules were kept. In many of those cases it was rather small changes that caused the troubles, because the previously used design principle had a very low factor of safety against damage; the small change transferred the structure from the "durable side" to the "non-durable". Some examples will be given:

* Concrete floor slabs in one-family houses were built directly on moist ground without changing the design that was previously used - with fairly good results - for floors on well-drained ground. Thus, no heat insulation was used in the floor. Even a very elementary moisture mechanics analysis had shown that this design meant great risks of damage (decay, bad smell, loosening) in conjunction with certain types of organic flooring materials. A change in RH from perhaps 90% in a well-drained ground to 100% in a moist ground is often disastrous. But such analyses were never made by the designers and the material producers did not point out the risks, simply because they did not realize that the environment in which their material should function was radically changed. Consequently, thousands of one-family houses in Sweden are now more or less damaged by moisture, due to a design based upon old, but irrelevant, experience.

* For concrete, the concept "quality" has normally been synonymous with the strength. Thus, in a more severe environment a higher strength level has been required. Strength has been used as a substituting property for W/C ratio which is in turn a measure of such properties as permeability, amount of freezable water, OH concentration of the pore water which effects the protection of the reinforcement, etc.

  The use of strength as a quality criterion is based upon the assumption of a relative well-defined relation between strength and W/C ratio. During the last decades there has been a gradual change in this relation, however, towards higher strength levels for the same W/C ratio; the cement strength has increased and the ready-mix concrete industry has become more skilled. Consequently, the same requirement with regard to strength means a higher W/C ratio and a lower durability today than it did 20 years ago. The mean relations for Swedish ready-mix concrete in the beginning of the 1960's and to-day are shown in Fig. 2. The change is rather drastic since even a small change in W/C ratio might have a very great effect on the durability.

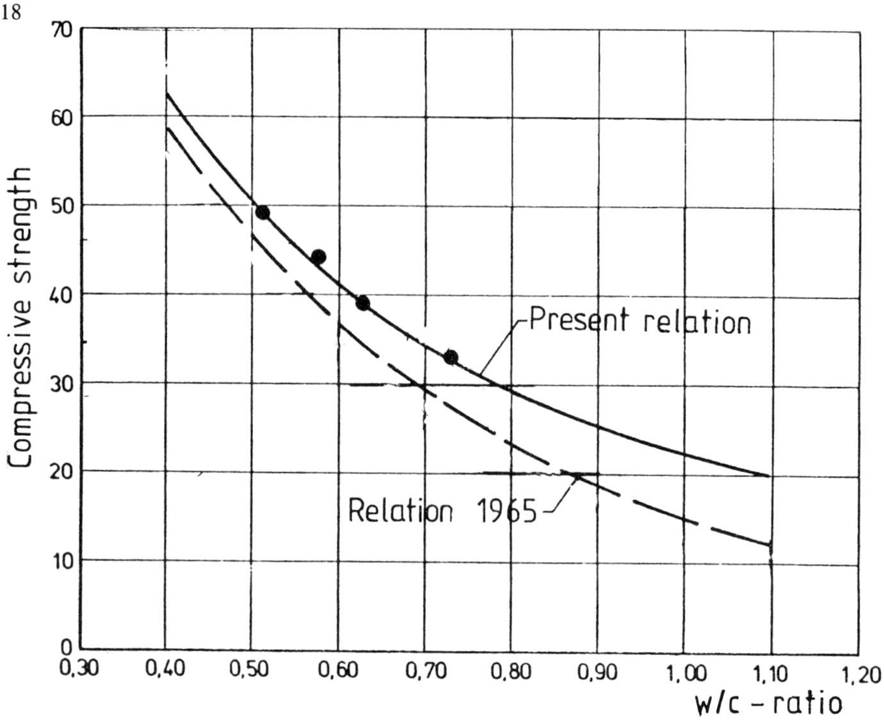

Fig. 2 The mean relation between water-cement ratio and compressive strength at 28 days for Swedish ready-mix concrete.

Similar cases can also be expected to occur in the future. Some Swedish examples of current changes in building design, in building materials and in environment that will have a great, but so far unexplored, effect on the service life of buildings will be described below. Some of them are certainly of current interest in many more countries.

* The transition from concrete floors directly on the ground to well insulated wooden floors separated from the ground by an aerated space. The risk of moisture induced damage in such structures is imminent unless the choice of materials and design are made with great care. This choice must be based on detailed analyses of the microclimate in the floor, and on functional analyses and tests of the durability of the materials used.

* Increased use of glued wooden fiber beams in highly insulated walls, floors and roofs instead of solid wooden girders. The reason for this is, of course, to reduce the building costs but it may also result in reduced durability. The effect of moisture and temperature on the mechanical and physical behaviour of that type of building element has not been studied very much.

* Additional heat insulation of older buildings. This means a change of the moisture state in the old part of the structure, eventually leading to incipient decay. Relatively small changes in the climate could be disastrous.

* Utilization of the possibility of heat accumulation in heavy building components; e.g. in thick concrete slabs on the ground. This will lead to new temperature and moisture fields that might reduce the service life if precautions are not taken in the design.

* Utilization of industrial wastes in the production of building materials, especially the use of different fly-ashes in cement and concrete. The effect on strength is often marginal and can be easily compensated for. The effect on durability, on the other hand, could be drastic. There are no reasons to believe that the "normal" relation between strength and durability should hold true. It is often claimed that materials such as fly-ash from coal burning or silica-fume should reduce the water permeability at equal strength and that this should be favourable for durability. This is nonsense because water permeability is only one parameter affecting durability. Such parameters as the pH-value of the pore water, base reserve, air content, air-pore structure, gas diffusivity are often influenced in a negative manner by the use of fly-ashes. Besides, concrete with fly-ashes are much more sensitive to bad curing than are normal concretes. Consequently, one cannot be sure that the low permeability of laboratory specimens will be reproduced in the surface layer of the real structure.

* Energy saving in the burning of ceramic products - tiles, bricks. This could reduce the frost-resistance and the chemical resistance; e.g. the resistance against alkali attack from the mortar. Old experience could not be relied upon in such cases.

* Increased acidification of the environment, which might affect the service life of underground structures - foundations, pipes etc.

All those examples show that one cannot rely upon experience or upon qualitative assessments of durability. It is quite clear that we must try to find quantitative methods of analyzing durability, preferably expressed in years of expected service life calculated during the design stage. It is then necessary to analyze not only each material separately but also the time-dependent interaction between all materials in the building component.

In the same manner, it is important that the service life concept be regarded more consciously in official standards and norms. The authorities must be aware of the fact that the fixation of a certain minimum quality level (e.g. the required strength or W/C-ratio of concrete in a certain environment) means that they have also settled the average service life of the whole future population of that type of concrete structure. Todáy such quality levels are settled on the rather defective information gained by "old experience" and unreliable simplified laboratory tests.

3 THE CONCEPT "SERVICE LIFE" - DEFINITION AND FACTORS TO CONSIDER

The service life of a building material or a building component is the space of time during which it fulfils all functional requirements. The concept service life is therefore meaningless unless it is possible to express the functional requirements in a perceptible, quantitative way. Consequently, considerable effort must be devoted to expressing the functional requirement in terms of measurable properties such as loadbearing capacity, deformation, strength, surface spalling, discoloration, emission of dangerous substances etc. Such functional requirements, although quantitative, are often very complex and composed of a number of more basic properties; e.g. the loadbearing capacity of a reinforced concrete beam exposed to frost attack is determined by such time-dependent factors as, (i) the long-term compressive strength of the concrete, (ii) the spalling depth of the concrete surface, (iii) the corrosion depth of the reinforcement bar /3/. All these basic properties must be analyzed if a service life estimate is to be possible. In the case of a reinforced concrete beam, therefore, at least three different deterioration mechanisms must be known /3/.

In many cases it is enough to study the interaction between the material or component and the external environment. In other cases it is also necessary to investigate the internal interaction between different materials. One example is the system "concrete on ground - putty - glue - carpet" which must be treated as a unity; the service life - e.g. the degradation of bond - could be quite different for different material combinations despite the fact that the external climate is constant.

The service life of a given structure is of course a function of the environment, as is illustrated by the following two examples:

Example 1;

Fig. 3 shows the degradation of bond between a plastic carpet and a concrete slab on the ground. The functional requirement is a bond of at least 0.2 MPa. This value is reached after

1 month when RH = 95% at the interface between the concrete and the carpet. It is reached after 3 months when RH = 90% but it is never reached when RH = 85%. In many practical cases, a high RH prevails only during a few months before the concrete slab has dried downwards towards a harmless equilibrium moisture content. Therefore no problems occur. In other cases the equilibrium moisture is high enough to give a low service life, or the time to reach a low equilibrium RH is so long that moisture-induced damage has time to develop.

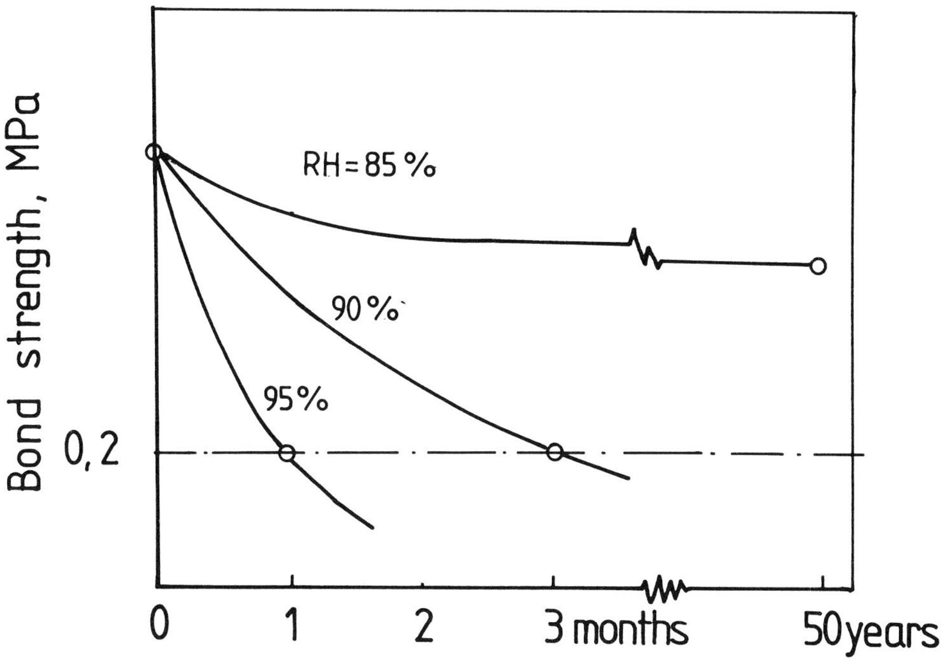

Fig. 3 Hypothetical curves over the degradation of bond between the carpet and the concrete slab.

Example 2;

Fig. 4 shows an outdoor parking deck in two different environments. The destruction curves are indicated. Environment 1 is completely "free of salts" while environment 2 contains both de-icing salts and freeze/thaw cycles. Loadbearing capacity is the function under consideration. In environment 1, no damage occurs until the carbonation front has reached the reinforcement bars (point A) at which point these start to corrode with a speed that is controlled by the oxygen diffusion into the bars. At a certain stage (point B) the concrete cover is spalled off and the function fails. The increase in serviceability until point A depends on the continued hydration. The service life is t_1.

In environment 2, a certain spalling occurs already during the first freeze/thaw cycles due to the de-icing salts. After one or two years (point C) chlorides of a dangerous concentration have reached the reinforcement bars at which point the passivation fails and the bars start to corrode. The serviceability diminishes right from the start; at first as a consequense of the spalling of the surface; after point C mainly due to corrosion. The minimum acceptable loadbearing capacity is already reached before the concrete cover is spalled off (which occurs at point D) as a consequence of the reduction of concrete and reinforcement cross-sections. The service life is t_2 which is much shorter than t_1.

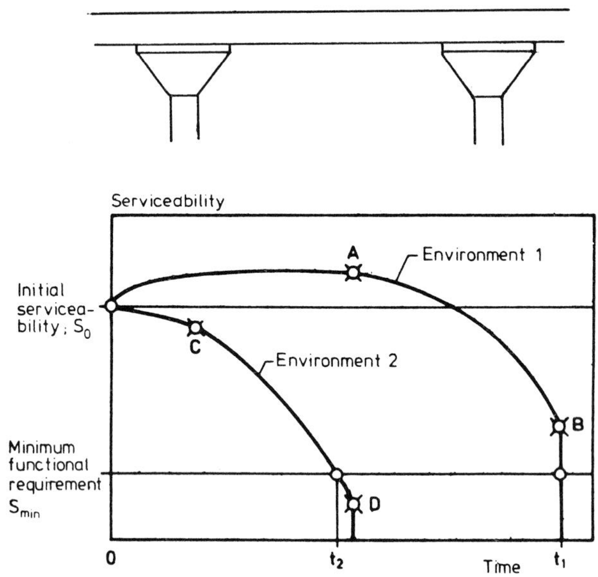

Fig. 4 Hypothetical serviceability-time curves of an outdoor parking deck. o indicates a change in the deterioration mechanism; see the text.

Some important (and some trivial) conclusions can be drawn from these two examples;

1: Different types of materials, e.g. different types of flooring materials, will have different serviceability - time curves in the same environment. The reason could very well be that quite different destruction mechanisms are acting. By utilizing the concept "service life" it is, however, in theory at least, possible to make a fair judgement between the different materials. Materials with the same service life have equal durability despite the fact that they are different in many other aspects, chemically and physically. Such a conclusion cannot be drawn on the basis of information from ordinary durability tests. These tests are often accelerated - see below - and quite different types of tests are often used for different types of materials. The interpretation problems are therefore considerable.

2: The functional requirement(s) must be clear and quantified. In Example 1 bond strength was used as the criterion for function. Another, equally important (or more important) criterion is "bad smell" which might appear long before the bond fails. This is caused by the decay of organic substances beneath the plastic carpet. Decay will follow other time-curves for each value of RH. Consequently, the service life will be different also.

The breaking down of a functional requirement into basic, measurable properties is often very complicated but is necessary for a reliable service life calculation.

3: The service life is highly dependent on the properties of the environment. This must therefore be known in detail - RH, temperature, precipitation, concentration of aggressive media etc. Possible future changes in the environment must also be considered; for example the recent use of large quantities of de-icing salts on concrete structures that were once designed for normal frost attack has created overwhelming problems on bridges in many countries; severe scaling, reinforcement corrosion.

4: The main destruction mechanism is often different during different periods. For example in Example 2 (Fig. 4), at point C, there is a change-over from frost attack to reinforcement corrosion as the most important destructive mechanism. Consequently, the serviceability - time curve changes its slope drastically. Such jumps between two mechanisms frequently occur in practice and must therefore be known in conjunction with the service life design. They also imply that one can seldom base a service life calculation upon simplified durability tests in which one type of destruction at a time is studied. Results of such tests - even if extrapolated correctly in time - will only yield a value of the service life of the actual specimen under the actual test conditions, but not of the real structure in the real environment. The complex, often synergetic, action of many simultaneous destructions might be analyzed theoretically on the basis of results of separate tests of each destruction, or on the basis of knowledge of the destruction mechanisms.

5: A service life prediction can only be made when the destruction mechanism is known in detail. For example, the induction period before the chloride corrosion starts in the parking deck in Fig. 4 is a function of numerous parameters connected in an intricate manner, which are still partly unknown. The basic properties are the OH^--concentration of the capillary water in the concrete and the diffusivity of Cl^--ions in the concrete cover /4/. But these two parameters are a function of many other properties; the W/C ratio, the cement content, the cement type, the type and amount of residual materials, the curing, the temperature etc. A mistake in the assessment of the effect of one of those properties will lead to great errors in the estimated service life.

6: Example 1, Fig. 3, shows that it is very useful if one single fundamental property could be found - in this case RH - which determines the degradation curve as well as the environment. This makes it much more easy to make a service life estimate. In order to find this fundamental property it is of course, as was mentioned above, necessary to know the destruction mechanism. Frost destruction of porous materials is an example of a durability problem that could be described sucessfully by one single fundamental property; viz the degree of water saturation S of the pore system. Each material has a certain critical (maximum allowable) value of S and each environment imposes a certain actual value of S in the material. The frost damage is a function of the probability that S_{ACT} exceeds S_{CR} /6/. The fact that S_{CR} exists is also predicted by the two most plausible destruction mechanisms, /6/.

So far, the serviceability - time curves have been assumed to be continuous; i.e. the destruction increases gradually with time. However, for some types of attack, the curves are discontinuous and severe destruction occurs more or less instantanously after a time of almost constant serviceability. A typical example of this feature is frost destruction; see Fig. 5.

The serviceability in the representative unit volume V in the material is constant or grows as long as the environmental conditions are such that the critical degree of saturation S_{CR} is not transgressed in the volume. S_{CR} is almost constant with time. Then, suddenly, at point B, the external conditions are very wet and S_{ACT} exceeds S_{CR} in conjunction with a freezing period. The destruction in the unit volume is then very large /5/ and its serviceability immediately drops to a low value. A real material consists of numerous unit volumes (size 1 -> 5 mm^3) for which reason destruction of one volume will not necessarily affect the material very much. Normally, however, a very large number of volumes are damaged simultanously, and a measurable drop in the serviceability of the material is therefore obtained.

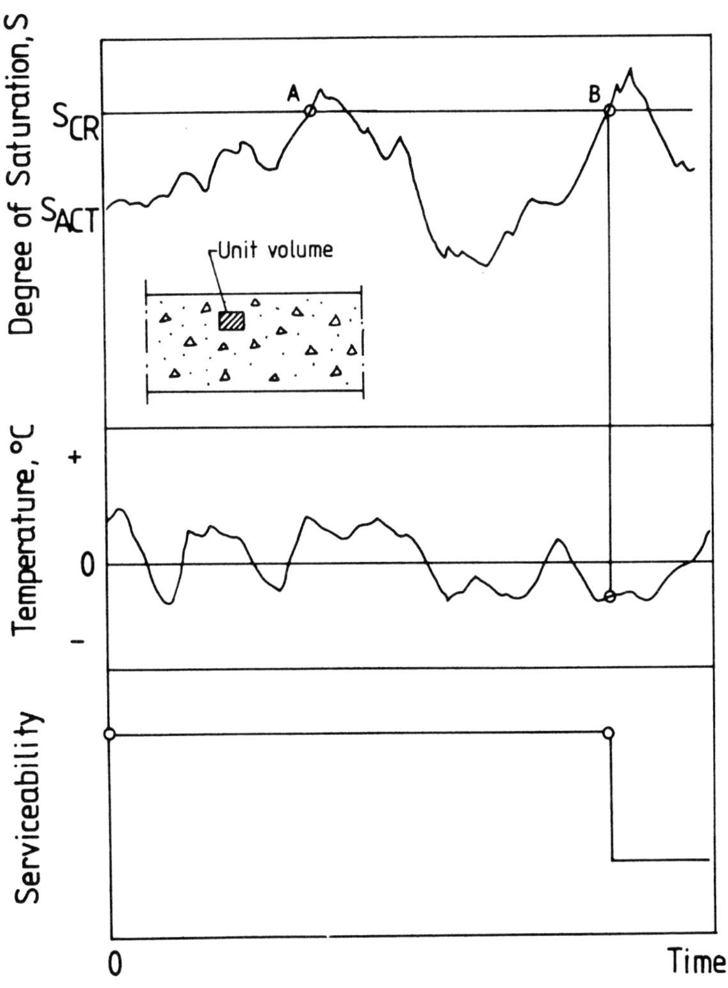

Fig. 5 Example of a discontinuous serviceability-time curve. The frost attack on a unit volume inside a porous material.

The frost destruction is a good example of the statistical character of service life. The very unlucky combination of too high a moisture content over a large material volume and freezing temperatures might occur as easily in the first year as in the fiftieth. Thus, it becomes very difficult to make a safe service life calculation. It is quite clear, however, that a long service life requires that the material or structure is designed in such a way that the probability that S_{ACT} should exceed S_{CR} becomes very low. I shall return to the frost attack below.

4 INFORMATION REQUIRED FOR A SERVICE LIFE PREDICTION

The information and methods that are needed for a reliable service life prediction can be summarized in the following way:

1: The functional requirements must be known and be expressed quantitatively in terms of one or more measurable basic material properties.

2: The environment around and inside the material, must be known as exactly as possible. It should also be expressed in quantitative terms. This part of the service life design requires, in almost all cases, a detailed analysis of the moisture-temperature field in the structure.

In most cases it is impossible to gain full knowledge of the environment; see Fig. 5 according to which one must predict the future moisture fluctations in all parts of the material in order to predict the risk of frost damage. Therefore, it is often necessary to employ some sort of standard environment which is intended to represent the real environment in a fair way. Then, one will not get the true service life but a potential service life; see below.

3: The interaction between the environment and the material - i.e. the destruction mechanism - must be known. A misconception concerning the destruction mechanism could lead to the use of a completely irrelevant durability test or to the stressing of an irrelevant material property. Such mistakes are frequent and examples are easy to find. Tests of frost resistance are often based upon the hypothesis that destruction is a consequence of fatigue. Therefore, a large number of freeze/thaw cycles are performed under uncontrolled moisture conditions. The gradual destruction is monitored. A typical example is the ASTM-test C 666 for concrete. Such a test will only give a qualitative value of frost resistance and cannot be used for a service life estimate. In reality, frost destruction is not caused by fatigue but by "single-load fracture" when freezing occurs at too high a moisture level. By utilizing the real mechanism, it is possible to design a test method by which the service life can be predicted, /6/. Another example is corrosion of reinforcement in a chlorine environment. Service life is often assumed to be a function of the diffusivity of Cl^--ions in the saturated concrete cover; a concrete with low diffusivity is considered to have a long service life. Therefore, the durability testing is often restricted to Cl-diffusivity only. In reality, however, there is a threshold concentration of Cl^--ions around the bar before corrosion can start. This concentration is very different for different concretes and has nothing to do with diffusivity. It is just as important as the diffusivity but is seldom measured.

4: All the information from points 1 - 3 must be merged together and used in the design of non-accelerated tests - see below - on the basis of which it is possible to assess the required serviceability - time curves after extrapolation of the results. Preferably, such tests should be excluded, however, in favour of theoretical models for describing the interaction between the material and the environment. The data to be put into the models are normally not based upon traditional durability tests but consist of structural, physical and mechanical properties of the material and of some environmental data. A theoretical calculation is considerably more rapid than an assessment based on durability tests. As examples of relatively elaborated theoretical service life models could be mentioned the reinforcement-corrosion model by Tuutti /4/ and the frost-attack model by the author /6/. As an example of a service life model based upon extrapolation of a single non-accelerated test could be mentioned the theory for acid attack on concrete elaborated by Rombén /7/.

The frost-destruction and the corrosion models are worked out in such great detail that it should, in theory, at least, be possible to calculate the service life on the basis of material and environmental data only; no durability test should be needed. The models are, however, not perfect. Many important parameters are still uncertain. Nevertheless, the models can already be used for a quantitative analysis of the effect on the expected service life of a change in different material properties or in the climate around the structure, see Fig. 6 and 7 below.

The requirements for a reliable service life prediction are clearly very high. This may be the reason for the limited use of the service life concept in building design. It is, however, quite clear that much more of the durability studies made must be directed towards destruction mechanisms and towards developing service life models. Today most studies unfortunately consist of rather rough investigations by accelerated durability tests yielding very little information of universal applicability.

5 THE POTENTIAL SERVICE LIFE

As has been mentioned above, it is very difficult to express the real environment in quantitative terms or to reproduce it in a non-accelerated test. Therefore, one often has to exchange the complicated real environment for a simpler "standard" environment which is easy to reproduce in the laboratory or to treat theoretically but which is still as similar to the real environment as possible.

By using the standard environment one will not get the true service life but merely a sort of "potential" service life. This concept has been used for frost attack /6/ and for reinforcement corrosion /4/. In the first case, the real condition has been exchanged for simple isothermal water absorption of a thin material slab put in contact with a free water surface. The potential service life is the space of time during which the degree of saturation reached during the capillary suction is lower than the critical value; see above. Thus, one has to extrapolate the measured absorption curve by a factor of 10 to 1000. An extrapolation of this magnitude can only be made if the mechanism is known by which the coarse pores in the material become water-filled. So far, the absorption has been assumed to be proportional to the logarithm of suction time. This is a considerable over-simplification and has in fact no physical significance.

Fig. 6 presents a chart for estimating the potential service life of concrete, exposed to freezing without salts. It is calculated theoretically on the basis of a destruction mechanism that leads to the existence of a so-called critical spacing between air-filled pores /6/. The air-void system - pores larger than 5 μm - is assumed to have the following frequency function of radii.

$$F(r) = k \cdot \frac{\ln 1.025}{1.025^{r}} \qquad \begin{array}{l} k = \text{constant} \\ r = \text{pore radius} \end{array}$$

The W/C-ratio is 0.60 or 0.45. The chart shows that a potential service life of 20 years ($1{,}7 \cdot 10^5$ hours of uninterrupted water storage) requires an air content of about 5.5% at a W/C-ratio of 0.45. The same air-content will only yield about 2 months of service life when the W/C-ratio is 0.60. The diagram is by no means exact due to difficulties in describing the water absorption process. Research is, however, now under way in order to clarify this problem.

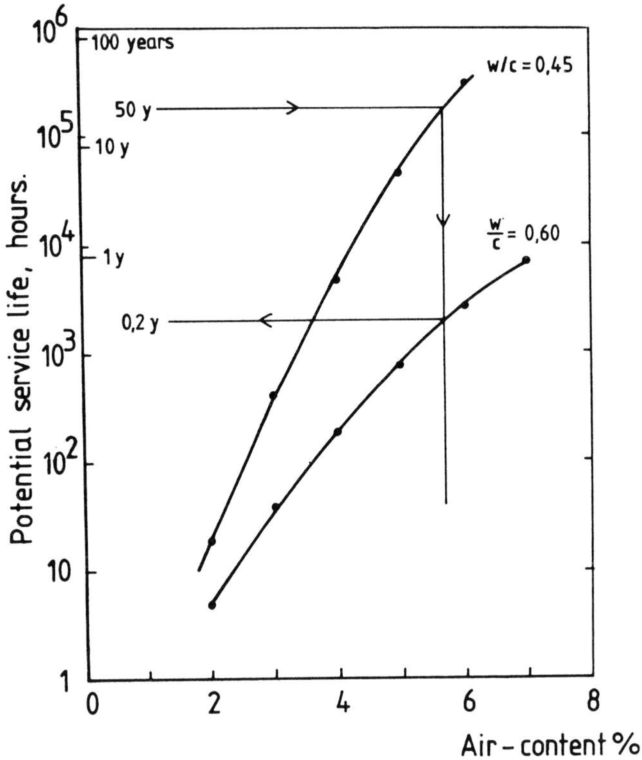

Fig. 6 The service life of Portland cement concrete subjected to uninterrupted water absorption followed by freezing.

In the case of reinforcement corrosion induced by carbonation of the concrete, the RH profile of the concrete cover is of major importance. The higher the RH, the lower the carbonation rate. Thus, a concrete which is constantly moist has a very low rate of carbonation and consequently a long service life. The same concrete placed outdoors but protected against rain has a rapid carbonation but is in most countries wet enough to have a rapid corrosion rate once the carbonation front has reached the bars. Thus, it has a low service life. The different types of environment can be divided in different classes, each with its own characteristics. Every class of environment will produce a certain RH profile in the concrete cover. It could be estimated by means of moisture mechanics analyses and will be different for different types of concrete depending on its W/C-ratio etc. On the basis of this information, it is possible to calculate the rate of penetration of the carbonation front /4/. Fig. 7 shows the results of such calculations valid for concretes with pure Portland cement and placed in two different Swedish environments; "outdoors protected from rain" and "outdoors exposed to rain". The figure is slightly adjusted in order to fit field data. It gives the maximum carbonation depths which can be expected. A W/C-ratio of 0.75 will only give a service life of 7 years for a rain-protected structure with a concrete cover of 20 mm. The W/C-ratio must be lowered to 0.50 if a service life of 50 years is required. The wetter structure has a service life of about 50 years and >100 years for those two W/C-ratios.

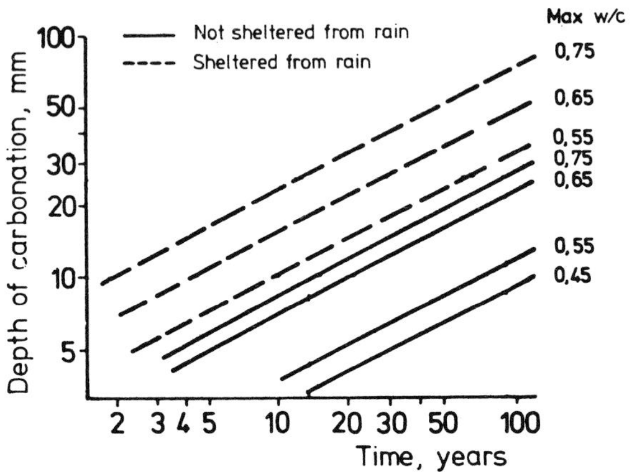

Fig. 7 Maximum carbonation depths of Portland cement concretes exposed to two different classes of environment /4/.

Tuutti /4/ has treated chloride-induced corrosion in a similar manner. Fig. 8 shows the result of his calculations. It gives the service life for the start of corrosion as a function of the diffusivity of Cl^--ions in the concrete cover and the threshold concentration of Cl^- in order to start corrosion. Those two factors are functions of the composition and curing of the concrete /4/.

The concept "potential service life" could probably be applied with success to many other types of destruction processes even outside the concrete field.

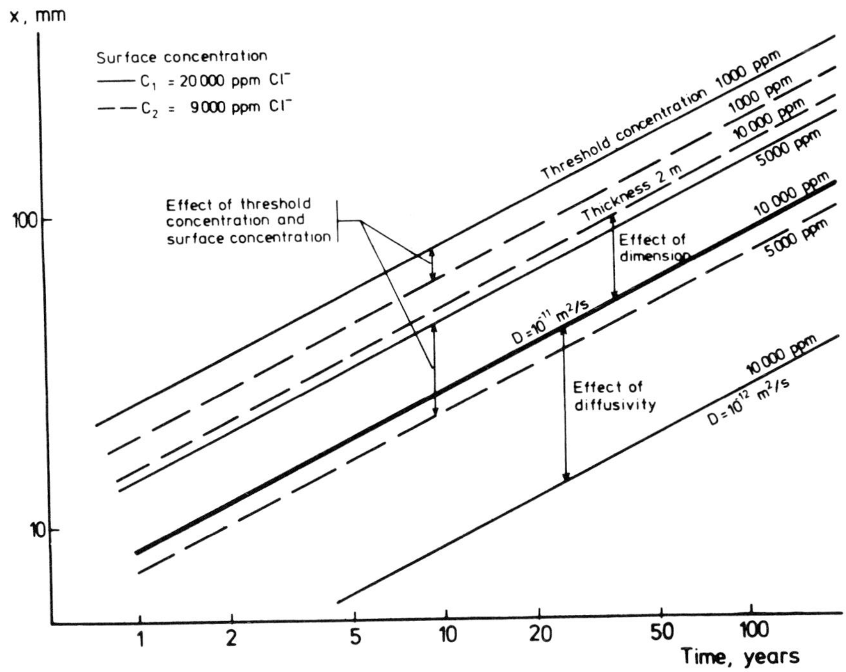

Fig. 8 The "incubation time" before start of reinforcement corrosion as a function of the Cl^--diffusivity and the threshold Cl^--concentration /4/.

6 INDIVIDUAL STRUCTURES VERSUS POPULATION OF STRUCTURES

The service life of one particular structure depends on the individual characteristics of exactly that structure and of its environment. All those characteristics have certain discrete values which, if known, make it possible, in principle at least, to calculate a discrete value of the service life of the actual structure, assuming that we know the destruction mechanism. The principle is shown in Fig. 9 (a).

In practice, however, we will never get access to all the information needed in order to calculate the exact value of the service life; all properties are afflicted with uncertainties. Therefore, we can only expect to predict the service life of the actual structure with a certain degree of uncertainty. However, if we know the mean value and the spread in all relevant properties it should be possible to estimate also the mean value and spread in the service life; see Fig. 8 (b). Such stochastical methods for concrete structures have been eleborated by Sentler /8/ among others.

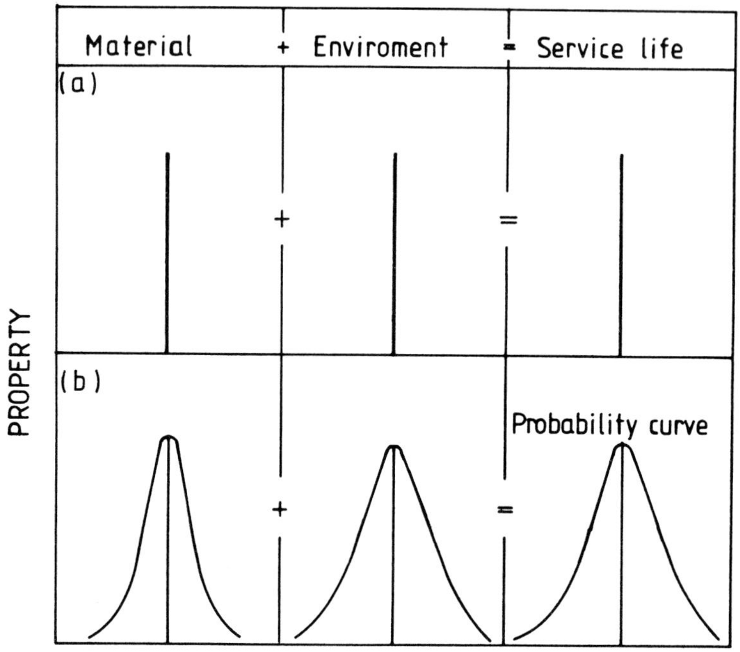

Fig. 9 The service life of a given structure.
(a) Exact data is available for material and environmental properties.
(b) Only the range of variation of all properties are available.

The same principles could be used for a whole population of similar building structures; e.g. all road bridges to be built in a country. In this case we have a much wider spread in material and environmental properties in connection with which parameters such as "quality" of the building work and climate during the building phase must also be considered since they considerably affect the properties of the finished building. By stochastical methods it should then be possible to estimate a frequency function of the service life for the whole population in which all parameters are considered.

Stochastical methods of this type ought to be used by authorities in connection with the establishment of quality requirements for different types of structures. This should result in considerably more rational requirements than the normal ones which are based upon "old experience" without any service life assessment.

Hypothetical examples of calculated frequency functions for two types of concrete structures; bridges and external walls - are shown in Fig. 10. For bridges, a characteristic service life of 100 years is required by the authorities, which means that only 5% of all bridges are allowed to have a service life shorter than 100 years. In order to achieve this, it is necessary according to the calculation to select concrete properties and design that give an average life of 150 years; a lower concrete quality will theoretically give an unacceptably low service life for a large fraction of all bridges. Similarly, for external walls, the required average life will be 75 years if the required characteristic value of 50 years should be achieved.

It is highly probable that the use of such methods has resulted in much higher quality requirements than those used in the majority of concrete standards in the world.

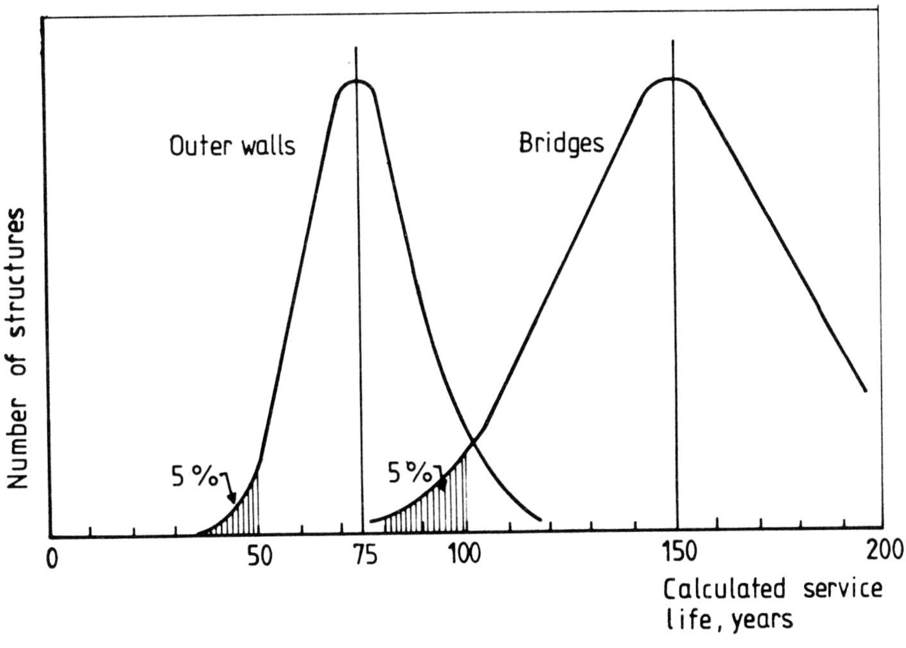

Fig. 10 Hypothetical frequency functions of service life for external concrete walls and concrete bridges. The curves are calculated on the basis of certain assumed material and environmental properties.

7 A "SENSITIVITY ANALYSIS"

It is characteristic for the durability problem that very small changes in the material properties or the design can cause problems and substantial reductions in service life. In many cases, it is easily neglected details in the design that turn out to be decisive for the service life; for example, the exchange after some years of the original flooring material for another type which has a lower moisture diffusivity or which has a lower critical RH - see Fig. 3 - might create moisture-induced damage in floors on the ground because the moisture state in the floor is changed - see Fig. 1.

One should of course not build houses that are that sensitive also to very small and natural changes. Therefore, the design should include a sensitivity analysis in which the effect on serviceability of imaginable future changes in the design is considered as well as the effect of maltreatment of the materials during the building phase, uncertainties in material properties and future accidents (e.g. water leakage).

A sensitivity analysis is nothing else but a service life analysis based on different assumptions. Materials for which relevant information is missing or is uncertain will be "punished" since they are either rejected or require a much more expensive design in order to be used safely. Thus, the use of sensitivity analyses of service life will hopefully stimulate the manufacturers to produce essential data.

8 DURABILITY TESTS AS A BASIS FOR SERVICE LIFE PREDICTIONS

It has already been said - Chapter 4 Subsection 4 - that the prediction of service life requires either a purely <u>theoretical model</u> for describing the interaction environment-material or some sort of <u>durability test</u> of the material or component followed by a translation of the test result into the expected service life.

Durability testing can be done according to two different principles:

1: <u>Accelerated testing</u> in which the environment is concentrated in such a way that the whole service life is run through during a short time in the laboratory.

2: <u>Non-accelerated testing</u> in which the actual environment is transferred unchanged into the laboratory. The material changes, which are normally very small, are monitored by sensitive methods, whereupon the whole destruction curve is extrapolated.

Accelerated tests are widely used, but are very unsuitable as a basis for service life predictions for the following reasons; see Fig. 11.

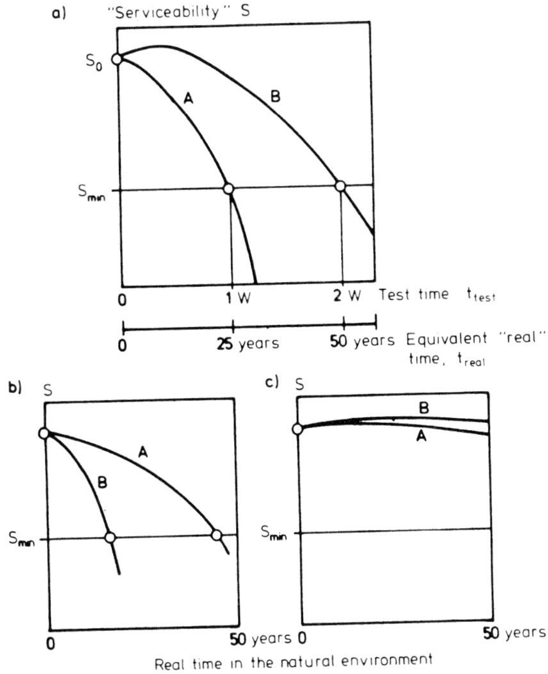

Fig. 11 Problems with accelerated tests
(a) The time scale
(b) In reality; A is better than B. In the test; B is better than A.
(c) In reality; both materials are highly durable. In the test; both materials are damaged.

1: It is almost always impossible to translate the exposure time in the test to a real exposure time in the real environment; in Fig. 11 (a) the time scale is supposed to be 1 week for each 25 years.

2: The acceleration sometimes changes the destruction mechanism. Consequently, it might very well be that material A is better than B in reality - Fig. 11 (b) - although, it was inferior in the test - Fig. 11 (a).

3: The acceleration is sometimes so large that destruction occurs although it would never happen in practice - Fig. 11 (c). In such a case the test is completely misleading.

4: In reality, one often has a synergetic effect between two or more destructive actions or time-dependent jumps between one action to another - see Fig. 4 above. This will not be revealed in a normal accelerated test since only one destruction type at a time is studied.

Non-accelerated tests can be used for service life predictions. This makes them very attractive: The problems with such tests are

1: Selection of the test environment which implies a careful analysis of the aggressive parts of the real environment.

2: Detection of the damage which is always very low due to the non-accelerated conditions and the short testing time.

3: Extrapolation of the measured destruction - time curve by a time factor 10^2 to 10^3.

The extrapolation problem is the most severe one because it requires full knowledge of the destruction mechanisms. One must also know about all mechanism changes that might occur in the future. A typical example of the use of a non-accelerated test for extrapolation of service life is the method elaborated by Rombén for assessing the service life of concrete pipes exposed to acid attack /7/. Rombén makes a thorough analysis of the destruction mechanisms and shows that the rate-determining process changes on two occasions. He shows that the service life will be considerably over-estimated if the last mechanism change occurring after some months is not considered in the extrapolation.

9 CONCLUSIONS

The present situation in which the building design with regard to durability is often based upon old experience and defective data concerning materials and environment is not desirable. It has been shown by many examples above that this design procedure could lead to considerable over-estimations of the durability. Besides, it does not stimulate the material producers to investigate their materials thoroughly or the designer to analyse the effect of a new design with regard to durability.

It would be much better if the qualitative concept "durability" was abandoned in favour of the quantitative concept "service life". Thus, each design should include a service life prediction. A safe service life prediction requires a lot of data, however; the functional requirements and the environment must be expressed in quantitative terms; the destruction mechanism must be known in detail; the interaction between environment and material must be calculated theoretically or extrapolated by a non-accelerated test, which means that a lot of material data must also be known.

The difficulties in making an accurate service life prediction are great. Examples from concrete research do, however, show that those difficulties might be coped with successfully if more durability research is directed towards really developing models for a theoretical description of the deterioration mechanism.

It has been shown that the real service life can often be exchanged for a potential service life which is valid for the component subjected to a reproducible standard environment which in turn is intended to represent the real environment in a satisfactory way.

The service life concept should always be used in connection with the elaboration of official standards in which the minimum quality levels of different types of structures are determined. This ought to lead to a higher probability that the required service life of the actual population of structures is reached than does the usual, more arbitrary, way of determining quality requirements.

The effect of uncertain material or environmental data on the service life can always be analyzed by a "sensitivity" analysis. This will "punish" materials for which relevant data are missing and will therefore stimulate the material producers to measure and present data that are of importance to the designer.

REFERENCES

1. Fagerlund G, Floors on the Ground. (Stockholm; BPA Byggproduktion AB, Handl Nr 32, 1980). (In Swedish)

2. Fagerlund G, A Theoretical Analysis of a Test Method for Capillary Suction of Granular Materials. (Stockholm; Swed. Res. Inst. for Cement and Concrete, Report Nr 7726, 1977). (In Swedish)

3. Fagerlund G, Predicting the Service Life of Concrete Structures, in J.F. Young, Ed., Characterization and Performance Prediction of Cement and Concrete. (New York: Engineering Foundation, 1982), pp 187-198.

4. Tuutti K, Corrosion of Steel in Concrete. (Stockholm; Swed. Res. Inst. for Cement and Concrete, Research Fo 4:32, 1982).

5. Fagerlund G, The Critical Degree of Saturation Method of Assessing the Freeze/Thaw Durability of Concrete. Materials and Structures 10(1977):58 pp 217-230.

6. Fagerlund G,. Prediction of the Service Life of Concrete Exposed to Frost Action, in Studies on Concrete Technology (Stockholm; Swed. Res. Inst. for Cement and Concrete, 1979).

7. Rombén L, Aspects of Testing Methods for Acid Attacks on Concrete - Further Experiments. (Stockholm, Swed. Res. Inst. for Cement and Concrete, Research Fo 9:29, 1979).

8. Sentler L, Statistical Evaluation of the Durability of Concrete. (Lund; Div. of Build. Techn., Lund Inst. of Technology, 1982).

REPORT OF DISCUSSION GROUP ON INORGANIC MATERIALS

Dr. T. P. Tassios
National Technical University
Athens, Greece

Participants: Emine Caner
Daniel Dobruse
Göran Fagerlund
J. Ferry Borges
Torben Hansen
Gerard Litvan
Hakon Olaffson
Leslie Parrott
Sven Pihlajavaara, Reporter
C. D. Pomeroy
Peter Pratt, Reporter
Th. Tassios, Discussion Leader

INTRODUCTION

The inorganic building materials considered by the Group included:

Cement pastes, mortars, concrete, fibre-reinforced mortars, reinforced concrete, gypsum products

Stone, bricks, masonry

Lime products, and

Pozzolans.

The general principles of service life prediction used in advanced technologies are applicable to these materials and probabilistic. methods, inspection schemes and fracture mechanics have already been applied in some instances. There appear to be many similarities of these materials with the polymeric materials described in the paper by Eurin, Marechal and Copé (this Workshop), since concrete often has an unknown water/cement (w/c) ratio and air content, may contain additives, suffers thermo-mechanical stress during curing and poses difficulties of quality control.

2 COMMONALITIES WITH ADVANCED TECHNOLOGIES

The prediction of service life for inorganic materials in building technology has many features in common with that in the advanced technologies. There is a need:

1. To formulate the requirements. This must be done in terms of measurable material properties, recognizing that these may be interrelated and dependent upon a variety of subproperties. In building technology, these will include not only strength, but also appearance, texture and acoustic and thermal properties, for example.

2. To describe environmental actions. This must be done in both space and time, recognizing the importance of the geographical location, the season and the orientation of the building. The complexity is such that "representative" standard environments have been envisaged for service life prediction.

3. To develop quality assurance schemes. There is no simple test to determine the free-water/cement ratio in ready-mix concrete and the air void content and spacing can only be determined with great difficulty. To assure the quality of the material delivered to the site, in-situ testing methods must be developed. Ideally, the data from these tests should be suitable for modeling the behavior as outlined in 6 below. There was considerable interest in further developing the established quality assurance schemes for design, construction and use although full and regular inspection schemes, as in the aircraft industry, were not envisaged. In-situ monitoring for some categories of structures might be very valuable.

4. To determine mechanisms of degradation. The fundamental mechanisms of degradation which determine the response of materials to their environment and synergies between these mechanisms must be better understood in building technology if progress in service life prediction is to be made. Extrapolation of in-service behavior (non-accelerated testing) can be used for comparison with the results of accelerated tests in the laboratory, with a view to identifying the problems of accelerated testing. Accelerated tests are frequently unsatisfactory in the complex field of building technology but we have to live with them and perform and interpret them intelligently. They must be backed by the highest possible knowledge of the degradation mechanisms and by calibration against in-service performance records. The

Group encouraged a relative shift of resources towards basic research on degradation mechanisms rather than on empirical accelerated durability tests.

5. To develop planned maintenance schemes. The Group agreed that planned maintenance schemes for buildings should be laid down by the designer, together with some form of inspection involving in-situ tests. Ideally, an owner's manual would be required which should be handed on at change of ownership. A reassessment of the predicted service life might be made on the basis of the inspection.

6. To develop mathematical models of the behavior of buildings and building materials. This must be based upon an adequate understanding of the mechanisms of degradation and of their synergies. This understanding will develop with time as a result of further research and the models must be reviewed and updated accordingly. If the models enable successful predictions to be made, even in a limited area, then communication between the materials scientist and the building engineer can be established on a basis of mutual confidence. Furthermore, in view of the complexity of the environmental action on building materials, the use of models can help to determine the relative weights of the different parameters and, therefore, become a tool for quantitative sensitivity analyses. In this way, modeling can help to identify the areas where basic research is needed, to stimulate manufacturers to provide the basic data required for prediction and to provide a better basis for simplified rules.

   Environmental loading is already recognized as probabilistic in nature in many branches of engineering and the Group expressed the hope that a "response spectrum" technique might be developed. A wide range of possible levels of environments with a certain probability of occurrence would be used parametrically as an input to the building material or building element system. In each case, the response of the system to the input would be determined as a function of a basic parameter of the system. In this context, the "response" would be a measure of a property after an environmental or in-service exposure (carbonation depth, degree of spalling, residual strength, etc.) and the "basic parameter" would be a selected critical value of a property or attribute (w/c ratio, air entrainment, permeability, etc.) which might describe the behavior of the system. The envelope of these "response" curves, the "response spectrum" could then be used to aid design. This technique appears to lie between the empirical approach to durability in the Codes and a detailed scientific analysis of each and every case.

3 DISSIMILARITIES WITH ADVANCED TECHNOLOGIES

A number of factors make the adoption of the techniques used by the advanced technologies difficult. Some of these have already been mentioned in section 2 of this report. These include:

1. The complex nature of the inorganic materials used in buildings and structures and the consequent difficulty in selecting a basic parameter which might determine their behavior.

2. The complex nature of their exposure environments.

3. The wide range of degradation mechanisms and the fact that these mechanisms are not yet fully understood.

In addition to these, there are the historical factors connected with the building industry:

4. The scale of the structures is often very large.

5. The lifetime of the structures is frequently long and owners and users may change.

6. Because it is widely distributed, the building sector is more comparable to a craft community than are the advanced technology industries.

7. The level of education of the <u>average</u> designer and contractor is probably lower than in advanced technologies.

These additional factors compound the legitimate variabilities due to the production process itself (e.g., the variability in concrete along the length of a vertically cast column) and make the development of quality assurance schemes all the more important. Unfortunately, Nature and the owners of buildings may have different values.

Probably the greatest difficulty arises from the corrosive nature of the environment. In the advanced technologies, the greatest successes in service life prediction have been due to the application of fracture mechanics to describe the growth of single, crack-like defects into unstable running cracks under an applied load. In building technology, multiple cracks are the rule and their growth is controlled more by chemical, physical and biological stresses than by mechanical stresses. Under these conditions, even the most sophisticated fracture mechanics analyses are invalid. Also, the nuclear and aircraft industries take great care to protect their products from corrosion by the environment.

4 ALTERNATIVE APPROACHES

In the long term, the need is to understand the basic mechanisms of degradation, to develop the science of fracture (rather than just the mechanisms of fracture) so as to include the mechanisms of degradation specifically and to intensify the education of structural engineers in durability subjects.

In the shorter term, the Codes of Practice require development and extension.

1. The Code rules should be scientifically justified as far as possible, instead of being presented as empirical facts.

2. The Codes must be designed so as to anticipate technological changes before they occur. It should be possible to reassess the durability of existing structures on the basis of appropriate inspections and developed models.

3. An Appendix should be provided in Codes describing the most recent scientific approach to service life prediction, even if it is difficult to apply in practice.

4. Higher safety factors are required for durability than at present. Whereas the safety factor on strength is typically 3, that on durability is nearer to 1.

5. Significant parameters have to be selected, as a result of both theory and experimental observation, for use in the Codes so that research on the one hand and construction rules on the other can develop side by side.

5 ADDITIONAL RISKS FROM USE OF NEW MATERIALS

New materials exposed to new environments frequently give rise to additional risks. To overcome these, more careful assessment may be required with more in-situ testing of a type less destructive than coring. Examples include the use of blended cements and the use of reduced temperatures in brick kilns.

The contact zone between different types of materials also poses additional risks and complexities for service life predictions. Examples include not only the steel to concrete bond, but also the problem of water proofing and insulating coatings trapping water or heat within load-bearing building elements, exposing them to frost damage.

REPORT OF DISCUSSION GROUP ON POLYMERIC MATERIALS

Mst. Dr. Ir. J. M. Rigo

Universite de Liege
Liege, Belgium

Participants: Erik Brandt
Robert Copé
Philippe Eurin
Tore Gjelsvik
Jonathan Martin, Reporter
Larry Masters
Lars Ostlund
J. M. Rigo, Discussion Leader
M. Fiebrich
Christer Sjöström

1 INTRODUCTION

That there is a growing use of plastics in applications that require careful design is undeniable; that this growth rate is as high as desirable is questionable. A major barrier to more extensive use of plastics is the lack of confidence of end users in the reliability of polymers. This stems from the fact that accurate data on the service life of polymeric products exposed to different service environments are not usually available to aid materials selection and use.

A major source of the design data that are available is the raw material manufacturer. These data alone, however, are not enough. They often inspire in the user an unwarranted level of confidence since it is often unclear how these data correspond to performance in the field. A clear understanding of the properties of plastics is essential to ensure that design data are relevant and this understanding must be supported by knowledge of how to apply these data to resolve problems in a cost-effective manner. For example, the consideration of cost-effectiveness in the selection and use of plastics requires knowledge of the service life of a product when it is subjected to its intended operating environment.

Important differences exist in the approach used by the building and construction technology and that used by advanced technologies in obtaining the data used in making selections between different materials. The building and construction industry is typically rather conservative in its selection of materials, relying on a performance listing of a product. This conservative approach is very resistant to the selection of new, innovative materials.

In order to obtain knowledge on the service behavior of a material, the building and construction industry relies heavily upon comparative data, in which the performance of a new material is compared to that of a "known" material under identical test conditions, and upon engineering judgment. Thus, service life is frequently addressed in a qualitative manner. The approach to generating service life data in advanced technologies is different in that heavy reliance is often placed upon accelerated test data for quantitative service life prediction. In the advanced technologies, the objectives of accelerated aging tests are:

1) To quantitatively predict the service life of a material or product;

2) To select the best product for a given application; and

3) To isolate and identify the causes and mechanisms of early degradation.

Implicit in these objectives is an understanding of the interrelationships among:

1. The performance properties of a product;

2. Changes in the chemical and physical properties of the materials as a function of time; and

3. The stress factors to which the product is exposed.

2 THE COMMONALITIES OF PROBLEMS

Despite the different approaches to service life prediction noted above, the Discussion Group noted a number of commonalities between service life prediction of polymeric materials in advanced technologies and in building and construction technology.

The discussion centered around the following generic steps, which relate to service life prediction:

1. Definition of requirements

   - Expected life time
   - Environment
   - Level of performance

2. Identification of available materials or products

3. Outdoor exposure to help identify degradation mechanisms

4. Accelerated aging tests to generate data on rate of degradation

5. Development of mathematical models

6. Design of the system and

7. Feedback of in-service performance.

2.1 Definition of Requirements

2.1.1 Expected life time. The required life time must be defined prior to the generation or evaluation of service life data. For example, the following table summarizes service life expectations resulting from an inquiry of specialists and users of buildings in Belgium.

Material or Component	Expected life time (years)
Internal Divisions	
Floor surfacing	15
Ceiling surfacing	40
Wall surfacing	35
Stairs	53
Division walls	55
Doors	38
Ironmongery	32
External Envelope	
Doors	42
Windows	45
Window shutters	35
Joints	35
Light fronts	48
Roof gutters	34
Lantern light	32
Roofs	45

One notes the typically very long expected service lives encountered in building and construction technology. Most often, these life times are defined as an average or mean service life; such a definition leads to difficulties in measurement. It is essential to define service life in such a way as to permit its measurement by accelerated tests. Advanced technologies on the other hand, usually prefer the definition of a minimum service life which is more easily measured using probabilistic principles.

2.1.2 Environment. An understanding of the environment in which the material or product is to be used is essential to predicting its in-service behavior. The aggressive agents of the environment can be classified in three families:

1) Physicochemical agents (humidity, temperature, light ...);
2) Physical agents (frost, hail, wind ...); and
3) Chemical agents (impurities in air ...).

Climate data are usually available for major cities and may include agents such as those listed below:

- Air temperature
- Soil temperature
- Relative humidity
- Precipitation
- Solar radiation
  - Sunshine duration
  - Direct radiation
  - Radiation on an horizontal surface
  - Radiative balance
  - Degree of cooling
- Wind
  - Velocity
  - Direction
- Water evaporation
- Atmospheric pollution
  - Radioactivity
  - Fumes
  - Sulfur dioxide
  - Atmospheric pressure

Most of these data are obtained hourly. These are usually rather complete and somewhat useful in service life prediction studies. Yet, these data are not sufficient to study the degradation mechanisms of polymers. For example, knowledge of the spectral distribution of sunlight may be much more relevant to polymeric research than knowledge of the total solar intensity. Also, the micro-climate may be significantly different than the climatic factors measured near major cities; the rate of degradation is dependent upon these micro-climatic factors and micro-climatic data are needed for service life predictions.

In approaching the determination of the effect of environment on materials and products, it must be realized that it is not possible to fully simulate the combined effects of all environmental factors in the laboratory. A possible alternative would be to identify major factors of importance and to systematically study the effect of these factors.

2.1.3 Level of performance. The user's requirements must be expressed in terms of a minimum acceptable level of performance, based upon essential properties. The properties used can frequently be considered in groups, such as aesthetic, comfort, and functional. The important point to make is that the defined properties must be expressed quantitatively for service life prediction. This need is true regardless of the material or product or the technological application.

2.2 Available Materials or Products

For both advanced technologies and the building and construction industries, one of the main problems with the use of polymeric materials is the great number and variety available. For example polyester, epoxy, polyurethane, polyvinylchloride, and polyethylene are all families of polymers in which one can find a large number of different formulations.

Further, it must be emphasized that the composition of a plastic product is frequently unknown. This points up the need for a strong data base on the performance of backbone polymers for use in studying modified polymeric materials and products. Also needed are data on the mechanical properties of polymers, reflecting the different behavior as a function of temperature or other environmental factors.

2.3 Outdoor Exposure to Help Identify Degradation Mechanisms

Exposure in the natural or service environment is important for studying and confirming mechanisms of degradation. Degradation can involve numerous processes, such as thermal, mechanical, photodegradation, degradation by high energy radiation, biodegradation, chemical decomposition, or can involve a combination of these processes. The possible resulting degradation mechanisms are often very complex. While natural exposures are often viewed as providing more reliable weathering data for service life predictions than accelerated testing, such exposures are usually very expensive in both time and manpower. Hence, it is important to use evaluation methods that can detect degradation in its early stages.

2.4 Accelerated Aging

The degradation of polymers, using simulated and accelerated exposure environments, offers a means of obtaining timely service life data. Accelerating degradation can typically be achieved by:

- an increase in concentration or in intensity of an aggressive agent or
- continuous exposure.

Essential to the development of reliable accelerated tests are knowledge of the degradation mechanisms and knowledge of the in-service exposure environment. Accelerated tests are particularly valuable in detecting early failures and it is for this purpose that they are extensively used in advanced technologies. For both advanced technologies and the building and construction technology, the need for reliable accelerated tests is great.

2.5 Mathematical Models and Statistical Analysis

Probabilistic methods are widely used in the development of mathematical models in the advanced technologies but are not widely used in assessing the service life of polymers in building and construction technology. The need for reliable models is common to all technologies in which service life data are needed.

2.6 Design of the System

One must always be cognizant of the size effect in the design of a system. For economic reasons, experiments are often conducted on small specimens. The results for these specimens, however, might be very different from those of the full-size products. Typically, the results of small specimens are much larger than those of the full-size specimens.

2.7 Feedback

Feedback of in-service performance data is carried out much more effectively in advanced technologies than in building and construction technology. Such data are essential to assessing the effectiveness of selection and use criteria and test methods for generating service life data. A need exists, particularly in the building and construction industry, for developing and implementing an effective method of feedback data.

REPORT OF DISCUSSION GROUP ON METALLIC MATERIALS

Dr. Steen Rostam
COWIconsult
Virum, Denmark

Participants: Frank Adams
Hans Arup
Erik Brandt
E. G. C. Coppens
Svein Haagenrud
Christopher Page
Steen Rostam, Discussion Leader
Sam Saunders
Gerhart Schuëller, Reporter
Ton Vrouwenvelder

INTRODUCTION

A discussion on the applicability of service life predictions within the building and construction sector should take into account the following three typical conditions:

- A small part of a structure may fail, e.g., concrete cover, cladding, etc., but the overall safety and stability of the structure is preserved (partial failure).

- Major parts of a structure or the whole structure may collapse without warning. This may be due to gross errors, which most often originate from the construction process.

- Two basically different conditions regarding ownership prevail in the building and construction sector:

  1) The structure is owned and operated by one single responsible owner from "cradle to grave" - the

"one-owner condition". This may often be by the state or by large state-like organizations and the typical examples of such structures are nuclear power plants, off-shore structures and bridges, or

2) The structure is owned and operated by multiple successive private owners with a very short horizon regarding economic involvement - the "multiple-owner condition". This is the most usual case for all ordinary dwellings and factory-like structures operated under a private or capitalistic economy.

It is important to realize that the large majority of structures and economic involvements in society are established and operated under the latter situation.

1 SERVICE LIFE PREDICTION

Scientists and engineers tend to treat service life solely as a technical problem, the solution of which leads to a technical service life of the structure. However, the economic aspects needed to establish and operate the structure will, in practice, be the decisive factor determining the economic service life of the structure. The service life is viewed ideally as the time, after construction, before the structure becomes technically - or structurally - obsolete. However, in practice, the usefulness of the structure may perish long before the technical or economic service life has been reached. The structure may become functionally obsolete, e.g., because allowable loads or required clearances have been increased.

Economic service life prediction - or life cycle cost optimization must include all operating costs including maintenance, repair and replacement, together with the initial costs for design and construction. Such a procedure is used in high technology industries and also for several building structures operated under the one-owner condition. However, the ordinary building and construction society is reluctant to accept this procedure because it feels that a service life prediction should be based mainly on conditions of the initial design concept and construction procedures.

Considering the type of problems involved in determining the service life of structures, a rational approach must be based on probabilistic methods in order to be truly operational in practice.

It is a fact that service life is a random variable. The aim of a service life design is to establish a target service life for the structure or type of structure considered. A mean service life or a design service life is not considered a clear concept to apply. The target service life concept includes the conditioned service life accounting for the modified target service life established for existing structures, based on accumulated experience from, for example, systematic inspections.

By the design, execution and foreseen upkeep, the target service life should be met with a certain probability, whereby service life is determined by the theories of reliability. Target service life is, thus, the span of time the structure fulfills its functional requirements without excessive cost for upkeep (maintenance).

For ordinary privately-owned type structures, maintenance cannot be fully relied upon and one procedure is to design and construct a robust structural skeleton and rely on the codes or specifications to ensure overall safety for the required time. For non-structural elements, a shorter service life may be acceptable, and maybe even desirable, as valuable modernization may result from such shorter service life of these elements.

A consequence should be that repair and modernization should be easy to perform, and hidden installations or the like should be totally avoided.

The concepts of conditional service life or residual service life of an existing structure, is a valuable application of the service life concept to determine frequency of maintenance, time intervals between inspections, planning for major repairs and for the whole economic planning performed by the owner. In the short term, this may constitute the most direct and most valuable application of the service life concept in the building and construction sector.

2 COMMONALITIES

Several commonalities were identified between problems and procedures in the advanced technologies and those encountered in building/construction technology.

Several procedures of the so-called "Advanced Technologies" seem to be applicable in special purpose structures, especially where fatigue and crack growth, partly combined with corrosion, are determining factors, e.g., off-shore structures, nuclear power plants and tall chimneys. Some of these procedures are already being utilized today in building/construction technology.

The discussion focused on the following items:

2.1 Force Management

An operation manual, analagous to manuals used in the advanced technologies (as aerospace vehicles), could also be employed for structures, e.g., in the form of an Operational Owner's Manual supplied by the contractor to the owner/operator of the structure, instructing in detail how the finished structure can be effectively serviced and maintained.

An example from Japan was reported in which the contractor not only bids for the erection of a structure, but also for the combined expenditures for construction and maintenance for a substantial period of time. Such a procedure is highly motivational in promoting high quality execution. This is an important factor in the building and construction industry, as the majority of causes for structural malformation or durability problems probably originate from events which occurred during the erection and construction process. Motivation to ensure high quality of a structure may also be supported by long liability periods.

The opportunities in the aerospace industry to select, regardless of cost, highly resistant material compositions, e.g., corrosion resistant alloys, cannot be utilized in the same manner in the building sector where economic restraints are of a more serious nature. While problems of metallic corrosion have been widely reported in the building sector, fatigue and crack growth have been the major problems in the advanced technologies. There the problem of fatigue was very serious and incorporated a multitude of influencing factors. A 20-year or more strong effort to tackle this problem has resulted in a drastic decrease in the severity of the problems, reducing it to an operational means of monitoring, testing and handling the problem.

It was proposed that the very important corrosion problems in the building sector - especially in chemical plants - be tackled in a similar way, applying a modified methodology concentrating on fewer numbers of critical parameters. Probability theory offers a means of tackling such problems in an operational way.

2.2 Individual Aircraft Tracking

In the areospace industry, the performance of individual aircraft is often tracked with increased time in service. Such tracking provides valuable data on the actual performance of aircraft and their components. Similar feedback data on buildings are seldom available. Participants felt that it is important, within the building sector, to generate feedback data but the data cannot likely be generated for all buildings because of the costs involved. One approach might be to generate representative data. For ordinary type structures, these data could be obtained for typical types of structures, such as dwellings, parking decks, bridges, factories, sports stadiums, etc.

The experience gained with a few selected types of behaviour for different structures of the same type would be valuable, not only for other similar structures, but also for the design of new structures.

2.3 Environmental Loading

The part of the environmental loading which comes from the atmosphere seems to be one of the clearest areas of commonalities between advanced technology structures and building structures as "we all live in the same environment".

The most important parameters are typically temperature, moisture and aggressive agents. In this respect, the micro-environment is rate-determining. Environmental characterization, including recordings of the micro-environment, its aggressivity, its relation to the macro or climatic environment and the classification of such environments in relation to aggressivity towards different building materials and components, is one of the most important subjects to study. This seems to be an area where the building sector may profit from ongoing activities in the advanced fields of technology.

In this respect, the building sector may profit from a more systematic approach to its problems by identifying the few most important parameters and then begin by concentrating on them. This would be valuable if we wish to develop equivalent design procedures when designing for physical, mechanical and for chemical loading. These procedures seem valuable from a design point of view and also seem highly feasible.

2.4 Damage Tolerance Approach

The damage tolerance approach, in which an initial reliability analysis is based on an assumption of the resistance to a certain maximum flaw (minimum detectable flaw) in the most adverse position, seems at present to be applicable only in special building structures such as off-shore structures and nuclear power plants.

For application to ordinary type structures, it would require a complete detailed structural survey at the "inauguration" of the structure, combined with a very detailed analysis, both of which are neither feasible nor necessary for such structures.

One first attempt along these lines, though, seems to be the development and use of continuously recording corrosion sensors now being installed in some buildings.

2.5 Prófit From Experience

The value of utilizing experience gained cannot be overemphasized. This requires an efficient recording of the experience and an efficient systematic procedure of feeding back information from "lessons learned", not only from major malfunctions, as disasters, but also from the systematic maintenance records of a structure.

The aerospace, nuclear and off-shore industries have developed systematics that may assist the building sector to develop an adapted systematics for incorporating feedback information into the future design and construction procedures. Moreover, the procedures could help in conditional service life evaluation and the modification of ongoing inspection and maintenance procedures.

The highway and bridge sector is one area within the building and construction field where rational strategies for such systematic inspection and maintenance routines already have been favorably employed for years.

2.6 Education and Information Transfer

The one most important basic element in preserving and improving the quality of structures and their performance is to ensure efficient and continuing education where new theories, new technologies and experience gained can be spread to a sufficiently large number of persons involved in the design, construction and upkeep of building structures.

The aerospace industry utilizes a "Damage Tolerance Handbook," where updated information on the theories applied and detailed information on all techniques available are kept available for the personnel. A loose-leaf document enables rapid and continuous updating of information and instructions.

In the building and construction sector, a Quality Assurance Manual would be the equivalent document, though the scope and level of detailing may differ.

Transferring relevant and updated information to the persons involved in decision-making and construction is acknowledged as the best means of ensuring high quality because it reduces faults and errors due to neglect or lack of knowledge. Not even the most strict control procedures can compensate for a lack of personal motivation to produce a good and reliable product, it being an aerospace vehicle or a building.

IV. MATHEMATICAL ANALYSIS TECHNIQUES USED IN ADVANCED TECHNOLOGIES

SOME MATHEMATICAL MODELS USEFUL IN RELIABILITY ANALYSIS FOR ADVANCED TECHNOLOGIES

Sam C. Saunders

Washington State University
Pullman, WA 99164-2930 USA

ABSTRACT

This paper presents the mathematical foundations thought necessary to formalize properly most instances of stochastic variation that will be problematic in the construction industry. An emphasis is placed on clarifying those aspects which are frequently misunderstood. The paper reviews the concept of a distribution, its sample and parameter spaces, and compares several of those that have proved to be most useful in life and strength calculations. The problems of estimating the somewhat specialized quantities of engineering interest are discussed. The modern methods of maximum likelihood estimation for incomplete samples are explained and illustrated. The classification of a life length by the behavior of its hazard rate is shown. Cumulative damage is defined and such distributions are given and their properties explained. Some problems from the aero-space and (Hi-tech) Electronic industries are presented to illustrate the combination of these concepts arising in actual problems.

1 FORMALIZING SCATTER AND VARIATION

Scatter in measurement and variation in physical resistance are an ubiquitous experience, and a common concern, to all engineers. The basic concepts that are necessary to treat such occurrences quantitatively, and correctly, are found in probability and statistics. The most basic ideas are based upon physical intuition in the same way that the ideas of point and line were undefined by the Greek geometers, but presumed to be clear to the Greek surveyor.

The Experiment

Consider an experiment, call it ξ, which may be regarded empirically as a set of laboratory procedures, but is essentially undefined. There is a universal set of natural conditions, call it Ω, of which the experimenter is ignorant but if $\omega \varepsilon \Omega$ could be known the outcome of the experiment $\xi(\omega)$ might be exactly determinable. In a coin tossing experiment ω would include the initial rotational and translational energy of the tossed coin, the trajectory of the arc, and the coefficient of elasticity of the impact surface. But it is presumed inconvenient, or impossible, to determine $\omega \varepsilon \Omega$ prior to the completion of the experiment so that $\xi(\omega)$ could be predicted.

All random experiments must possess these properties:

a. Each experiment can, at least conceptually, be repeated indefinitely under essentially unchanged conditions.

b. Although we cannot state what any individual outcome will be, we are able to describe the set of all possible outcomes of the experiment.

c. As the experiment is performed repeatedly, the individual outcomes are unpredictable (irregular) but as the experiment is repeated a large number of times a pattern, or regularity, appears in the frequency of outcomes.

It is this global regularity which makes it possible to construct a precise mathematical model with which to analyze the experiments which correspond.

Sets and Events

Let S be the set of all possible outcomes of an experiment, i.e. nature chooses an $\omega \varepsilon \Omega$ and the experiment ξ is performed with outcome $\xi(\omega) \varepsilon S$. The set S is called the <u>sample space</u> of ξ, or <u>sure event</u>, which is presumed known.

Given a sample space S and the corresponding experiment ξ we say an event $A \subset S$ occurs if and only if $\xi(\omega) \varepsilon A$, i.e. the experimental outcome is a point in A. If A and B are events, further events can be formed by unions, intersections and complements and thus the calculus of mutually exclusive and/or exhaustive events becomes an exercise in Boolean Algebra.

Let ξ be an experiment, with sample space S. We repeat the experiment n times, say $(\xi(\omega_1), \xi(\omega_2),\ldots, \xi(\omega_n))$ with $\xi(\omega_i)$ being the ith repetition; the sample space for this combined experiment is the n-dimensional space $S \times S \times \ldots \times S$ with element $(s_1,\ldots, s_n)$ where each $s_i \in S$. In this repeated experiment we may want to ask questions about the relative frequency of an event A, call it $f_A(n)$, during n trials.

Now we note these properties from our assumptions concerning random experiments:

1. $0 \leq f_A \leq 1$,
2. $f_A = 1$ if and only if A occurs every time,
3. $f_A = 0$ if and only if A never occurs,
4. if A, B are mutually exclusive $f_{A \cup B} = f_A + f_B$,
5. in some sense $f_A(n)$ converges in the limit as $n \to \infty$ to the probability of A.

Note that all of these properties can be empirically verified.

Probabilities

Those persons who believe that probability can be determined without experimentation, by introspection, personal belief, hunches, insight or even engineering knowledge, proceed somewhat differently with their "analyses" than those who believe that observed frequency is a necessary prerequisite for construction of a mathematical model, as well as the ultimate test of its acceptability. Such beliefs are called personal probabilities.

The average of the personal probabilities of persons in the queue's at the para-mutual betting windows determine the odds on the board for wagering at the races. And those persons who believe their probability of success at the gaming table changes with luck are welcome universally.

Of course personal opinion should be included in any decision that affects only oneself. However, prudence dictates that such nonscientific methods, with their potential for self-serving bias, be avoided in analysis involving public interests.

From the empirical properties (1), (2), (3), (4) we postulate a function to measure probability, call it P, which satisfies the following.

Definition: For a given experiment ξ with a sample space S and any event $A \subset S$ associate a real number P(A), called the probability of A and:

1. $0 \leq P(A) \leq 1$.

2. $P(S) = 1$.

3. If A, B are mutually exclusive events, i.e. $AB = \emptyset$ then $P(A \cup B) = P(A) + P(B)$.

4. If $A_1, \ldots, A_n, \ldots$ are pairwise mutually exclusive events (i.e. $A_i A_j = \emptyset$ unless $i = j$) then $P(\bigcup_1^{\infty} A_i) = \sum_{i=1}^{\infty} P(A_i)$.

This definition sets out the formal relationships between probabilities, events and sample spaces which must apply in all models. But the actual construction of a probability distribution on a sample space, which is a sufficiently accurate model of a physical situation that realistic predictions can be made on the basis of the calculated probabilities of events, depends not only upon a knowledge of probability theory but even more upon a knowledge of the physical-(chemical) nature of the phenomenon being modelled.

Whenever a measurement, or count, is made on the sample point determined by the outcome of an experiment it is called a random variable (r.v.). For example, the experiment could be tossing a coin four times and observing the sequence of heads and tails; for which counting the number of faces is a random variable. We write

$$P[X \varepsilon A] = P[s \varepsilon S: X[\xi(s)] \varepsilon A] .$$

Again, injecting white mice with a disease, waiting a fixed period and then disecting them, is the experiment. Counting the leukocytes in the blood could be the random variable of interest.

If X_1 and X_2 are two measurements on the same experiment then two events A and B are independent iff

$$P(AB) \equiv P(X_1 \varepsilon A, X_2 \varepsilon B) = P(A) \cdot P(B);$$

the r.v.'s X_1 and X_2 are independent iff this equality holds for all events A and B. If r.v.'s are independent and identically distributed we write i.i.d.

Finite Sample Spaces and Discrete Random Variables

Let $S = \{x_1, \ldots, x_n\}$ contain only a finite number of elements. If S is the sample space of some experiment, then every singleton event $\{x_i\}$ is called an elementary event. Let $p_i = P\{x_i\}$ be the probability of the ith elementary event. Then we know that $p_i \geq 0$ for $i = 1, \ldots, N$ and $\sum_1^N p_i = 1$ and since any event $A \subset S$ is a collection of elementary events, $P(A) = \sum_{x_i \varepsilon A} p_i$. Very little more can be said until $(p_1, \ldots, p_N)$ are determined.

When the sample space is countable, say $S = \{x_1, \ldots, x_n\}$, or $S = \{x_1, x_2, \ldots\}$ the random variable is discrete, and the discrete probability density, is determined in accord with the model, say,

$$P[X = x_i] = p_i \quad \text{for} \quad i = 1, \ldots, N \quad (\text{or} \quad i = 1, 2, \ldots).$$

Here S is the sample space of X, the space of values into which X must fall.

One of the most useful discrete distributions is the Binomial. Assume a r.v. X counts the number of good items in a sequence of n trials which are independent, each trial having the same unknown probability p of being good. Let q = 1-p then

$$P[X = x] = \binom{n}{x} p^x q^{n-x} \quad \text{for} \quad x = 0, 1, \ldots, n$$

is a binomial density. A related discrete density is that of Pascal (geometric)

$$P[X = j] = q^{j-1} p \qquad j = 1, 2, \ldots \quad .$$

Here X counts the number of i.i.d. trials until the first success.

There is another distribution related to the binomial. If n is large and p is small then the following approximation holds

$$\binom{n}{k} p^k q^{n-k} \doteq \frac{e^{-np} (np)^k}{k!} \quad \text{for} \quad k = 0, 1, \ldots \quad .$$

Thus a r.v. X has the Poisson distribution with parameter $\lambda > 0$ whenever

$$P[X = k] = \frac{e^{-\lambda} \lambda^k}{k!} \qquad k = 0, 1, \ldots \quad .$$

Since, as will be shown later, λ is the mean value of X it must have the same dimensions. If λ were the expected number per unit time then for any time t we would expect the number λt to occur. The actual random number to occur in (0,t), say N_t, would also have a Poisson distribution

$$P[N_t = k] = \frac{e^{-\lambda t}(\lambda t)^k}{k!} \qquad k = 0, 1, 2, \ldots \quad .$$

If λ is the expected number per unit area, per unit volume, per sheet or per jar then the probabilities of the number of occurrences, when it has a Poisson distribution, can be calculated for different volumes, areas, etc.

Infinite Sample Spaces and Continuous Random Variables

Frequently, the r.v. X has as its sample space the entire real line i.e., $S = \mathbb{R}$. In this case the basic events are of the form

$$[\omega : X(\omega) \le x] \text{ usually written } [X \le x] \quad \text{for any} \quad x \in S.$$

The probability of this event, as a function of $x \in \mathbb{R}$, is the distribution of X, usually denoted by F, with the proper affix X, namely

$$F_X(x) = P[X \le x] \qquad \text{for} \quad -\infty < x < \infty \ .$$

In the case the function F_X is differentiable then a density $F_X' = f_X$ exists and is related to the distribution

$$X \sim F_X \quad \text{means} \quad F_X(x) = \int_{-\infty}^{x} f_X(y)dy \qquad \text{for } -\infty < x < \infty$$

and we say X is continuous in this case.

For any r.v. X, either discrete, continuous or mixed, with distribution F_X, written $X \sim F_X$, the expected value of any function $\psi(X)$ is given by

$$E\psi(X) = \int_{-\infty}^{\infty} \psi(x)\, dF_X(x) = \begin{cases} \sum_i \xi(x_i)\, p_i & \text{, if discrete} \\ \int_{-\infty}^{\infty} \psi(x)\, F_X'(x)dx & \text{, if continuous.} \end{cases}$$

The mean (expected value) and variance of X are, respectively, defined by

$$\mu = E(X) = \int_{-\infty}^{\infty} x d\, F_X(x)$$

$$\sigma^2 = E(X-\mu)^2 = \int_{-\infty}^{\infty} (x-\mu)^2\, dF_X(x)\ .$$

One of the most useful distributions is the Gaussian (or normal). Its density is the familiar, symmetric, bell-shaped curve given by

$$f(x) = \frac{1}{\sqrt{2\pi}} e^{-\frac{(1-\mu)^2}{2\sigma^2}} \qquad -\infty < x < \infty .$$

Here $\mu \in \mathbb{R}$ is the mean (location parameter) and $\sigma > 0$ is the standard deviation (scale parameters).

Random variables $X_1,\ldots,X_n$ are said to be independent if and only if

$$P[X_1 \leq x_1,\ldots, X_n \leq x] = \prod_1^n P[X_i \leq x_i] \quad \text{for each } x_i \in \mathbb{R}$$

or equivalently the joint density function factors into a product

$$f_{X_1,\ldots X_n}(x_1,\ldots,x_n) = \prod_1^n f_{X_i}(x_i) .$$

We have surreptitiously introduced the parameters of a distribution which we now may formalize by defining a family of distributions: A random variable X has the distribution $F(x:\theta)$ for some parameter $\theta \in \Theta$. It then becomes a problem of mathematical statistics to utilize the data from the random sample $(X_1\ldots,X_n)$ to estimate the particular value of θ which governs the distribution of X.

If X has a normal distribution with mean μ and variance σ^2 we write $X \sim N(\mu, \sigma^2)$. If $Z \sim N(0,1)$ we say it is a standard normal; it has tabled distribution Φ and density φ, see [1], by which any other Gaussian distribution can be expressed, e.g. if f_X is the density of X which is normal then

$$f_X(x) = \frac{1}{\sigma}\varphi\left(\frac{x-\mu}{\sigma}\right) \qquad \text{for } x \in \mathbb{R} .$$

One of the reasons for the usefulness of this distribution is the fact given a large number of independent r.v.'s, where X_i has mean μ_i and finite variance σ_1^2 for $i = 1,\ldots, n$, whether they have the same distribution or not, their sum $X_1 +\ldots + X_n$ will have approximately a $N(\mu_1 +\ldots +\mu_n; \sigma_1^2 +\ldots +\sigma_n^2)$ distribution. This implies that sample mean $\overline{X}$ of n i.i.d. observations each with mean μ and variance σ^2 will have an approximate $N(\mu; \sigma^2/n)$ distribution. N.B. If the X_i are normal then the sum is exactly normal.

This asymptotic result is called the Central Limit Theorem because it is the limit theorem central to so much statistical analysis. Moreover it is the basis for the assumption of normality in these cases where each measurement obtained is the sum of several independent causes.

A related distribution is called Galton's Law or the log-normal distribution.

If $T > 0$ is a time-of-failure with a log normal distribution this means that $X = \ln T$ (or $\log T$) will be normally distributed. Thus the density of T is given by

$$f_T(t) = \frac{1}{\sqrt{2\pi}\,\sigma\, t}\, e^{-\frac{1}{2}\left[\frac{\ln(t/\beta)}{\sigma}\right]^2} \quad \text{for} \quad t > 0.$$

This density is skewed to the right. If the parameter β is the characteristic life; the modal or most likely value, is ν and the mean is μ. We find $\nu < \beta < \mu$ with

$$\mu = ET = \beta e^{\sigma^2/2}, \quad ET^2 = \beta^2 e^{2\sigma^2} \quad \text{and} \quad \nu = \beta e^{-\sigma^2}.$$

The log-normal distribution is, from the Central Limit Theorem, the approximate distribution whenever the r.v. is the product of a number of positive independent factors.

Another family of distributions is called the gamma family:

$$X \sim \Gamma(\alpha,\beta) \text{ means } f(x) = \frac{x^{\alpha-1}\, e^{-x/\beta}}{\beta^{\alpha}\Gamma(\alpha)} \quad \text{for } x > 0\,.$$

Here

$$\Gamma(x) = \int_0^{\infty} t^{x-1} e^{-t} dt \quad \text{for any} \quad x > 0: \quad \text{note } \Gamma(n) = (n-1)!$$

The cumulative distribution is given by

$$F(x) = \int_0^x f(t)dt = \int_0^{x/\beta} \frac{t^{\alpha-1} e^{-t}}{\Gamma(\alpha)}\, dt\,,$$

the values of which can be obtained from tables, see [2] or from a machine program. By straightforward integration we find

$$EX = \alpha\beta \quad \text{and} \quad VAR(X)\ \alpha\beta^2.$$

This density is always unimodal and for $\alpha > 1$ is skewed to the right but for $\alpha < 1$ is unbounded and decreasing.

2 ESTIMATING AND DEMONSTRATING RELIABILITY

The objective of statistics is to make inferences about a population based on information contained in a sample. Since populations are characterized by their parameters (measurements on the population), the subject of many statistical investigations is to make inferences about one or more parameters. Most statistical procedures involve either estimation or hypothesis testing. We will be concerned only with the topic of estimation.

Estimation has many practical applications in engineering. The only unalterable requirement is to obtain a sample which represents the population without unknown bias. A random (I.I.d.) sample from a population is one in which confidence can be placed, although obtaining such a sample may be difficult.

Problem 1: We would like to determine the applied stress at which say, only 5% of a certain building component will break. Can we estimate the mean breaking strength of this fifth percentile from a set of data? Is it really necessary to fracture all the specimens, if this value is all that is desired?

Let S be the set of all such components that are, and will be, built. Let X denote the breaking strength of a component chosen at random from S. Let F denote the distribution of X. We want to find the value of stress ζ such that on average only one in twenty will fracture. We must solve for ζ in the equation $F(\zeta) = .05$ and then estimate it.

Suppose $X_1, \ldots, X_n$ is an i.i.d. sample of size n from S i.e. X_1 is the first component tested etc. Let $X_{1,n} < X_{2,n} < \ldots < X_{n,n}$ be the ordered sample values. To estimate ζ it is necessary only to determine $X_{k,n}$ where $k = [\![\frac{n}{20}]\!] + 1$ and $[\![x]\!]$ is the largest integer not exceeding $x > 0$. Often this can be done without fracturing many of the components.

Alternatively suppose that we know more, namely, that X has some distribution within a known parametric family $F(x : \theta)$ for $\theta \varepsilon \Theta$. Then we utilize the sample data to estimate the parameter, call the estimate $\hat{\theta}$, and then solve the equation $F(\zeta : \hat{\theta}) = p$ to obtain an estimate of ζ_p, the 100 p^{th} percentile.

For example, if we know the data is normal we estimate the unknown population mean and variance and use the estimates to calculate the 100 p^{th} percentile for any p.

Problem 2: Suppose we would like to determine the mean service life of a certain electrical component but at the present time only two out of the fifty in service have failed.

Let T be the (random) life of such a component. We want to find the expected life of the components i.e. the value μ such that

$$\mu = ET = \int_0^\infty t\, f(t)dt .$$

Denote the two failed observations by t_1, t_2 while the remaining components have served for times t_i for $i = 3,\ldots, 50$ without failure. Since we want the expected life, we can utilize the sample mean of these that failed, namely $(t_1 + t_2)/2$ as an estimate. But is this the best that can be done, ignoring the survival information in t_i, for $i = 3,\ldots,50$?

Note that in this case our estimate is based on intuition. We now look at a more formal approach for deriving estimators; namely, the method of maximum likelihood that can be applied to (virtually) any situation.

The technique called the method of maximum likelihood selects as estimates those values of the parameters that maximize the likelihood (probability) of the observed sample.

Theory 1:

Let $X_1, X_2,\ldots, X_n$ be i.i.d. sample from $f(x;\theta)$ where θ is unknown. The Likelihood Function is defined for the sample $\underline{x} = (x_1,\ldots,x_n)$ by

$$L(\theta|\underline{x}) = \prod_{i=1}^{n} f(x_i;\theta). \tag{1}$$

NOTE: The likelihood function gives the likelihood that the random variables assume a particular value. It is the value of a density function; so for discrete random variables, it is a probability.

We are interested in finding the value of θ which maximizes $L(\theta|\underline{x})$.

Definition: The value $\hat{\theta}$ which maximizes $L(\theta|\underline{x})$ is called the maximum likelihood estimator θ. (MLE)

NOTE: $L'(\theta|\underline{x})$ and $\ln L(\theta|\underline{x})$ have their maxima at the same value and often it is easier to find the maximum of $\ln L(\theta|\underline{x})$.

Example 1: Suppose a random sample of size n is drawn from the Bernoulli distribution, namely,

$$f(x;p) = p^x(1-p)^{1-x}, \; x = 0,1; \; 0 < p < 1.$$

The sample vlues $x_1, x_2, \ldots, x_n$ will be a sequence of 0's and 1's.

The likelihood function is given by

$$L(p|\underline{x}) = \prod_{i=1}^{n} p^{x_i}(1-p)^{1-x_i} = p^{\Sigma x_i}(1-p)^{n-\Sigma x_i}.$$

Using techniques of calculus one determines the value of p which maximizes $L(p|\underline{x})$; it is $\hat{p} = \frac{1}{n}\sum_1^n x_i$ not surprisingly.

Example 2: We cannot count the number of bacteria in water, we can only test whether or not any are present. Suppose the number present in a given volume of water v is Poisson with parameter λv.

Let X = # of bacteria. Then

$$f_X(x;\lambda) = \frac{e^{-\lambda v}(\lambda v)^x}{x!}, \; x = 0, 1, 2, \ldots .$$

We observe $y_i = I\{x_i = 0\}$ (i.e., we count the number of pure samples). Using the data $y_1, y_2, \ldots, y_n$, we estimate the bacterial density λ. Now $f(y) = p^y(1-p)^{1-y}$ where

$$p = P[Y_i = 1] = P[X_i = 0] = e^{-\lambda v}.$$

$$L(\lambda|\underline{y}) = \sum_{i=1}^{n} [y_i \ln p + (1-y_i)\ln(1-p)]$$

$$\frac{\partial L}{\partial \lambda} = \frac{\partial L}{\partial p} \cdot \frac{\partial p}{\partial \lambda} = \sum_{i=1}^{n} \left[\frac{y_i}{p} - \frac{1-y_i}{1-p}\right] \cdot e^{-\lambda v}(-v) = 0.$$ This implies $\hat{p} = \overline{y}$ and $\hat{\lambda} = \frac{1}{v}\ln(\overline{y})$.

In volume v_i, we have indicator y_i, i = 1, 2,..., n, of whether contaminated or not. We now consider the problem of estimation in the general case.

$\hat{\lambda}$ is the solution of the equation in the variable λ

$$\sum_{i=1}^{n} v_i = \sum_{i=1}^{n} \frac{(1-y_i)v_i}{1-e^{-\lambda v_i}} .$$

Here we must use some numerical techniques to solve for $\hat{\lambda}$.

Theory II:

Let $\underline{t} = (t_1, t_2, \ldots, t_k, t_{k+1}, \ldots, t_n)$ be a censored sample where

t_i, $i = 1,2,\ldots, k$ are dead times

t_j, $j = k+1,\ldots, n$ are alive times.

Then the likelihood function becomes, setting $R(t:\theta) = 1 - F(t;\theta)$

$$L(\theta|\underline{t}) = \prod_{i=1}^{k} f(t_i;\theta) \prod_{j=k+1}^{n} R(t_j;\theta). \tag{2}$$

Example 1: Exponential Distribution for $T > 0$.

Since

$$P[T > t] = R(t;\theta) = e^{-\theta t}$$

$$f(t) = \theta e^{-\theta t} , \; t > 0,$$

we find

$$L(\theta|\underline{t}) = \prod_{i=1}^{k} \theta e^{-\theta t_i} \prod_{j=k+1} e^{-\theta t_j} .$$

Hence we see the log-likelihood is

$\ln L = k \ln\theta + (-\theta \sum_{i=1}^{n} t_i)$ whence differentiation shows the MLE to be $\hat{\theta} = k/ \sum_{i=1}^{n} T_i$.

There are of course other methods of estimation. We mention two:

(1) The method of moments. The population moments, which contain the unknown parameters, are equated to an appropriate number of the sample moments and the resulting equations solved simultaneously. This requires complete samples and moreover seems to be of utility only when the equations are the same as the maximum likelihood equations, as they are in the normal case.

(2) Bayesian (Empirical Bayesian) estimation: Regard the unknown parameters as random variables across the sample space, the

distribution of which is constructed a priori from personal belief. Often, if the distribution is unknown the uniform distribution is chosen from a belief of equal ignorance. After the sampling is drawn a distribution is constructed a posteriori wherein the personal belief is modified by experience. In the empirical Bayes technique the prior distribution is constructed from data also.

Some of the desirable properties of the MLE's are as follows:

a. While MLE's are not in general unbiased, they can usually be modified slightly to make them so.

b. MLE's are invariant, i.e. if $\hat{\theta}$ is the MLE of $\theta \varepsilon \Omega$ and u is a 1-1 function on Ω then $u(\hat{\theta})$ is the MLE of $u(\theta)$.

c. If a parameter does have a sufficient estimator, i.e. it contains all the relevant information, then the MLE will be a sufficient estimator.

d. MLE's are asymptotically normally distributed with the smallest variance of any other estimator which is asymptotically normal. Hence MLE's are asymptotically optimal.

On the other hand maximum likelihood cannot be used unthinkingly. For example, if one takes the gamma distribution, $\Gamma(\alpha,\beta)$, with likelihood

$$L(\alpha,\beta|\underline{x}) = n\alpha \ln\beta + (\alpha - 1) \sum_1^n \ln x_i - \frac{1}{\beta}\sum_1^n x_i \, .$$

The solution of the stationary poin $\frac{\partial L}{\partial \alpha} = 0$, $\frac{\partial L}{\partial \beta} = 0$ is a MLE for $\alpha > 1$ but a minimum likelihood estimator for $\alpha < 1$. This is essentially because for $\alpha < 1$ we have a decreasing failure and the density is markedly different.

Part of the perceived difficulty with MLE's in practice was due to their existence only as implicit solutions (this is now obviated by computing machines) and the uncritical use of stationary points of the likelihood functions.

Property (b) will be of particular importance in the discussion which follows concerning survival distributions.

3 SURVIVAL AND RELIABILITY

In life or strength testing the sample space is usually $\mathbb{R}^+ = [0,\infty]$, since the measurement cannot be negative. If T is the random life length of a component, one is often interested in the survival distribution defined by

$$\bar{F}_T(t) = 1 - F_T(t) = P[T > t] \quad \text{for} \quad t \geq 0 .$$

It is called the reliability at time t, i.e., the probability of surviving more than t units of time. When one thinks of life length one is naturally interested in such things as the conditional probability of surviving a period of time h, knowing it has survived to age t, hence

$$P[T < t + h | T > t] = \frac{P[t < T < t + h]}{P[T > t]} = \frac{F_T(t+h) - F_T(t)}{\bar{F}_T(t)} .$$

The hazard rate (or force of mortality) at age $t > 0$ is, by the preceding equation

$$q(t) = \lim_{h \to 0} \frac{1}{h} P[T < t + h | T > t]$$

$$= \lim_{h \to 0} \frac{F(t+h) - F(t)}{h} \cdot \frac{1}{1 - F(t)} = \frac{f(t)}{1-F(t)} . \qquad (3)$$

The force of mortality is so intuitively appealing, that its conception predates that of density. Any function $q(t) > 0$ for $t > 0$ such that its integral diverges, i.e. $\int_0^\infty q(x)dx = \infty$, will serve as a hazard rate. To see this integrate the equation above to obtain the hazard function Q defined by

$$Q(x) = \int_0^x q(t)dt = \int_0^x \frac{f(t)}{1 - F(t)} dt = -\ln[1 - F(t)]. \qquad (4)$$

Rewriting we obtain

$$\bar{F}(t) = 1 - F(t) = e^{-Q(t)} = e^{-\int_0^t q(x)dx} \quad \text{for } t > 0.$$

If your hazard rate decreases (sometimes labeled DHR) the longer you live the healthier you are and the longer you are expected to survive. If your hazard rate increases (sometimes labeled 1HR) then the older you are the shorter is the time remaining.

Many living things, and many complex systems, exhibit a bathtub shaped hazard rate with infant mortality in the initial portion, a constant hazard rate during mid-life, when death is due to accident, and wearout when the hazard rate increases terminally.

This phenomenon gives rise to the concept of "burn-in" or "green-run" which is used to eliminate "infant mortality" or test for "manufacturing error" on many systems and components.

If T has hazard rate q then a "burn-in" of duration τ will result in a random life remaining T_τ with survival distribution

$$\bar{F}_{t_\tau}(t) = P[T > t + \tau | T > \tau] = e^{-\int_\tau^{t+\tau} q(x)dx}$$

If $T \sim F$, a safe life is the life length t_ϵ for $\epsilon > 0$ but very small, such that

$$P[T < t_\epsilon] = F(t_\epsilon) = \epsilon \ .$$

The value $1-\epsilon$ is the level of safety.

Example 1.

The hazard rate q(x) of the $\Gamma(\alpha,\beta)$ can be shown to be given by

$$\frac{1}{q(x)} = \int_0^\infty (1 + \frac{u}{x})^{\alpha-1} e^{-u/\beta}\, du \ .$$

Thus we see that q(x) is an increasing function of x if $\alpha > 1$ but decreasing for $\alpha < 1$. Moreover for all $\alpha > 0$, we find

$$\lim_{x\to\infty} \frac{1}{q(x)} = \int_0^\infty e^{-u/\beta}\, du = \beta \ .$$

But accordingly as $0 < \alpha < 1$, $\alpha = 1$ or $\alpha > 1$ we have a DHR, constant hazard rate, or IHR distribution.

Example 2: The normal distribution is IHR. The hazard rate q for the $N(\mu,\sigma^2)$ can be related to that of the standard normal say $q_0 = \varphi/(1 - \Phi)$ by the equation

$$q(x) = \frac{\exp[-\frac{1}{2}(\frac{(x-y)^2}{\sigma})^2]}{\int_x^\infty \exp[-\frac{1}{2}(\frac{t-\mu}{\sigma})^2]dt} = \frac{1}{\sigma} q_0(\frac{x-\mu}{\sigma})$$

and q_0 can be shown to be monotone increasing using Mill's Ratio

$$1 - \frac{1}{x^2} < \frac{x}{q_0(x)} < 1 \qquad \text{for} \quad x > 0 \ .$$

Example 3: The log-normal distribution has a failure rate which is unimodal and ultimately approaches zero. One checks that the hazard q can be written

$$q(t) = \frac{1}{t} q_0\left(\frac{\ln t - \ln\beta}{\sigma}\right) \qquad \text{for} \quad t > 0$$

where q_0 is the hazard rate for the N(0,1) distribution. Calculation proves the truth of the claim.

4 THE WEIBULL DISTRIBUTION

In 1939 a Swedish engineer, Wallodi Weibull, began to apply the "weakest link" theory to the strength of materials. In two papers in that year he made the assumption that $F_0(\sigma)$, the probability of breakage of a unit volume as a function of stress σ is given by the statistical distribution

$$F_0(\sigma) = 1 - \exp[-\sigma/\sigma_0)^m] \quad \text{for } \sigma > 0$$

where σ_0 and m are unknown parameters which depend on the characteristics of the material under test.

In 1951 Weibull published one of the most influential statistical papers of modern times, see [3]. The distribution which he advocated was known in statistics as the third asymptotic distribution of smallest values and has since become as important in engineering applications as the normal distribution is in other applications.

Example: Suppose X represents the strength of a chain of n links, and let X_i denote the strength of the ith link. The strength of the chain is equal to the strength of its weakest link, $X = \min(X_1, X_2, \ldots, X_n)$. Consequently the distribution of X is the distribution of a minimum. For many different types of random variables, the limiting distribution of the minimum approaches a Weibull distribution as $n \to \infty$. (In a traffic flow, a driver's speed is constrained by the slowest driver. Blood-flow in arteries is similarly constrained.)

Let T be a continuous nonnegative random variable; it has a Weibull distribution, $T \sim W(\alpha,\beta)$, if and only if it has pdf (density)

$$f(t) = \begin{cases} \frac{\alpha}{\beta}\left(\frac{t}{\beta}\right)^{\alpha-1} e^{-(t/\beta)^{\alpha}} & t > 0;\ \alpha,\ \beta > 0 \\ 0 & \text{otherwise,} \end{cases}$$

where α is a shape parameter and β is a scale parameter, often called the characteristic life.

NOTE: $E(T) = \beta\,\Gamma(\frac{\alpha+1}{\alpha})$ $Var(t) = \beta^2[\Gamma(\frac{a+2}{\alpha}) - \Gamma(\frac{\alpha+1}{\alpha})]$.

The reliability if T is a life-length, is given by

$$\overline{F}(t) = e^{-(t/\beta)^{\alpha}} = e^{-e^{\alpha(\ln t - \ln\beta)}} \quad t \geq 0$$

and the hazard rate by

$$q(t) = \frac{f(t)}{1 - F(t)} = \frac{\alpha}{\beta}\left(\frac{t}{\beta}\right)^{\alpha-1} \quad t \geq 0\ .$$

NOTE: q is decreasing for $\alpha < 1$, increasing for $\alpha > 1$ and constant for $\alpha = 1$.

One of the most useful applications of the Weibull distribution is to employ its extreme-value property. To illustrate: Let $T_1,\ldots,T_n \sim W(\alpha,\beta)$ and set $Y = \min(T_1,\ldots,T_n)$ then

$$P[Y > t] = P[T_1 > t,\ T_2 > t,\ldots,T_n > t] = \prod_{i=1}^{n} P[T_i > t]$$

and by the i.i.d. property

$$\left[e^{-(t/\beta)^{\alpha}}\right]^n = e^{-(t/\beta_1)^{\alpha}} \text{ where } \beta_1 = \frac{\beta}{n^{1/\alpha}}$$

thus we conclude that $Y \sim W(\alpha,\ \beta n^{-1/\alpha})$.

Example 1: Suppose β = 50,000 hrs., n = 2000 units and α = 5. By the above we see β_1 = 10,933.6 hrs.

We can expect the most likely time of the first failure will occur before 13,616.4 hrs. Other probabilities can be calculated using the Weibull Distribution.

Estimates of the parameters α and β of the Weibull distribution are somewhat difficult to obtain. We consider the case where we have an incomplete sample with k failures out of n data points: assume theory II, thus by the notation used in equation (3) one obtains:

$$\text{Ln } L(\alpha,\beta|\underline{t}) = k \ln\alpha - k\alpha \ln\beta + (\alpha-1)\sum_{i=1}^{k} \ln t_i - \sum_{i=1}^{n} (t_i/\beta)^{\alpha} ,$$

from which we find

$$\frac{\partial \ln L}{\alpha\beta} = \frac{k}{\alpha} - k \ln\beta + \sum_{i=1}^{k} \ln t_i - \sum_{i=1}^{n} (t_i/\beta)^{\alpha} \ln(t_i/\beta), \quad (5)$$

$$\frac{\partial \ln L}{\alpha\beta} = -\frac{k\alpha}{\beta} + \sum_{i=1}^{n} \frac{t_i^{\alpha}}{\beta^{\alpha+1}} . \quad (6)$$

Setting (5) and (6) both equal to 0, after simplification, yields two equations

$$\frac{1}{k}\sum_{i=1}^{n} (\frac{t_i}{\beta}) - \frac{1}{k}\sum_{i=1}^{k} \alpha \ln (\frac{t_i}{\beta}) = 1$$

and

$$\beta = \left(\frac{1}{k}\sum_{i=1}^{n} t_i^{\alpha}\right)^{1/\alpha} .$$

Define the monotone increasing function A by the equation

$$A(\alpha) = \frac{\sum_{i=1}^{n} t_i^{\alpha} \ln t_i}{\sum_{i=1}^{n} t_i^{\alpha}} - \frac{1}{k}\sum_{i=1}^{k} \ln t_i \quad (7)$$

and solve $A(\alpha) = \frac{1}{\alpha}$ using computer methods to obtain $\hat{\alpha}$. Then set

$$\hat{\beta} = \left(\frac{1}{k}\sum_{i=1}^{n} t_i^{\hat{\alpha}}\right)^{1/\hat{\alpha}} .$$

Once we have obtained these estimates we can use the invariance property of MLE's to estimate the reliability and hazard function, i.e.

$$\hat{R}(t) = e^{-(t/\hat{\beta})^{\hat{\alpha}}} , \quad \hat{q}(t) = \frac{\hat{\alpha}}{\hat{\beta}}\left(\frac{t}{\hat{\beta}}\right)^{\hat{\alpha}-1} \qquad t \geq 0 .$$

Example 3: Let us assume that the hazard function for a complex piece of equipment is the sum of three Weibull hazards, viz.,

$$Q(t) = \left(\frac{t}{\beta_1}\right)^{\frac{1}{2}} + \frac{t}{\beta_2} + \left(\frac{t}{\beta_3}\right)^4 .$$

Here the failure takes place at the minimum of three independent causes. These summands are, respectively, the hazard due to infant mortality resulting from a flaw in manufacturing or assembly; accidental failure or random failure due to some unassignable cause, and finally the hazard due to the wear-out or fatigue of components.

From these assumptions we obtain a so-called bath-tub hazard rate

$$q(t) = \frac{1}{2}\left(\frac{t}{\beta_1}\right)^{-\frac{1}{2}} + \frac{1}{\beta_2} + \frac{4}{\beta_3}\left(\frac{t}{\beta_3}\right)^3 .$$

As long as we can assign the cause of failure to each malfunction independent of the time of occurrence, then it is possible to estimate all these parameters from one set of failure data.

Suppose we record the failure times and the cause of failure in the vector

$$(t_{11},\ldots,t_{1,k_1},\ t_{21},\ldots,t_{2,k_2},\ t_{31},\ldots,t_{3,k_3},\ t_{41},\ldots,t_{4,k_4})$$

where t_{ij} indicates the ith cause for i = 1, 2, 3 and the jth failure due to that cause while t_{4j} are the run-outs or nonfailure times during that test.

Using the formulae derived perviously:

$$\hat{\beta}_1 = \left(\frac{1}{k_1}\sum_{i=1}^{4}\sum_{j=1}^{k_i}\sqrt{t_{ij}}\right)^2 ,$$

$$\hat{\beta}_2 = \frac{1}{k_2}\sum_{i=1}^{4}\sum_{j=1}^{k_i} t_{ij} ,$$

$$\hat{\beta}_3 = \left(\frac{1}{k_3}\sum_{i=1}^{4}\sum_{j=1}^{k_i} t_{ij}^{\,4}\right)^{\frac{1}{4}} .$$

If the corresponding shape parameters are unknown the problem becomes harder but it is still possible to obtain the requisite estimes by computer programs.

5 CUMULATIVE DAMAGE OR MAXIMUM LOAD

Let us consider a structural component which is repeatedly subjected to a duty cycle of periodic loads which contain oscillations of fluctuating magnitude. In each duty cycle the component sustains some incremental damage (or wear). The question is how long will the component last before it must be replaced (or serviced).

The first engineering rule for calculating the life under such regimes was derived by Palmgren (1924), rediscovered by Miner (1945), and so is called Miner's Rule of cumulative damage.

The assumptions usually made to obtain Miner's rule may be summarized as follows:

(a) The amount of damage absorbed by the material in any one oscillation is determined only by the load during that oscillation.

(b) Each specimen can absorb the same amount of damage, and when that amount is attained, failure occurs.

(c) The total damage absorbed by the specimen under a sequence of load oscillations is equal to the sum of damages absorbed in each oscillation during the sequence.

Miner's Rule: Under assumptions (a), (b), (c), if a given cycle contains n_i oscillations of load ℓ_i where N_i is a number of oscillations to failure under the same load for $i = 1,\ldots,k$ then the number of such cycles that can be repeated until failure is

$$N = 1/\sum_{i=1}^{k} \frac{n_i}{N_i} .$$

Hence, by the cumulative damage rule we determine the fraction of damage accrued during one cycle and use its reciprocal to estimate the total life. The average damage rate per cycle is the harmonic mean of the rates of accumulated damage for each oscillation in the spectrum of oscillations. But this deterministic rule, found not to be sufficiently accurate in many cases. See [6].

Suppose that the incremental damage is not constant during each duty cycle but is random. Let Y_i be the random incremental damage during the ith duty cycle for $i = 1,\ldots,n$. Then $S_n = Y_1 + \ldots + Y_n$ is the total cumulative damage at the end of n cycles. If W is the total cumulative damage which the structure can sustain before failure (and is itself random since each structure can withstand somewhat different amounts of total damage) then the number of cycles until failure will be a random variable. We must calculate its distribution.

Let N denote the number of cycles until failure takes place. It is random and is defined by the event

$$[N=n] = [S_{n-1} < W,\ S_n \geq W] \quad n = 1, 2, \ldots$$

Hence the probability distribution of N can be determined from the behavior of the random sums S_n for $n = 1, 2, \ldots$.

Suppose that the damage increment due to the ith duty cycle is itself the sum of oscillations, say $Y_j = X_{1,j} + \ldots + X_{m,j}$, for any j, of which exactly k are distinct, and that the ith distinct load oscillation call it l_i, occurs n_i times for $i = 1,\ldots, k$. Then $n = n_1 + \ldots + n_k$ and if μ_i is the expected number of oscillations until failure under repetitions of l_i then it can be proved under regular conditions on the distribution of the $X_{i,j}$'s that

$$\frac{1}{\sum_{i=1}^{k} \frac{n_i}{\mu_i}} - 1 \leq EN \leq \frac{1}{\sum_{1}^{k} \frac{n_i}{\mu_i + 1}} .$$

This says that Miner's rule is true only in expectation and then only when conditions are such that increments are independent with a hazard rate that increases on the average, see [8].

But to derive results that are more useful we must be more specific about the manner in which it occurs.

Model I Random Damage due to Accumulated Shocks

Now let us assume that damage does not occur regularly with each duty cycle but instead occurs as a result of shocks from loads that occur at random in time according to the Poisson process with rate λ. This is the case with gust loading for airplanes or truck loads on bridge girders. Assume that the ith shock will cause a random amount of damage say $Y_i > 0$ for $i = 1,2,\ldots$ When the critical amount of damage, say $W > 0$, has been accumulated from the repeated shocks the device will fail, or must be repaired.

Let F_n be the distribution of $S_n = Y_1 + \ldots + Y_n$. i.e. $P[S_n \le y] = F_n(y)$ for $y > 0$ and G the distribution of W, the initial strength of the material. Then the probability that the device does not fail before time t is given by the

$$\overline{H}(t) = \int_0^\infty \sum_{n=0}^{\infty} \frac{e^{-\lambda t}(\lambda t)^n}{n!} F_n(y)G'(y)dy.$$

This then is the reliability of a structure which is accumulating random amounts of damage at times which are themselves random according to a Poisson process.

Model II Random Shocks but Damage only at Overloads

Suppose that shocks occur at random according to a Poisson process but that the structure, like a tree in a windstorm, does not accumulate damage but will only suffer when the overload due to the shock is too great. Let us assume that for every time in the duty cycle of a structure there is a mean activity level, say a, and there is at this activity level a mean time λ_a between overloads due to shocks. Whenever the overload occurs at activity level a it has a magnitude which is random with destruction G_a. This behavior may be gust activity on an airplane at a given altitude during the ground-aid-ground cycle of an airplane or the diurnal traffic intensity of trucks on a bridge and the consequent stresses imposed on girders.

At a fixed activity level a, let $N_a(t)$ be the counting process for the random number of overloads which will occur in (0,t). Thus

$$Y_1, Y_2, \ldots, Y_{N_a(t)} \quad \text{for } t > 0$$

are the random magnitudes of the overloads. Let us consider the maximum overload in time interval [0,t], call it $X_a(t)$. We may calculate its distribution by

$$P[X_a(t) > y] = \sum_{n=0}^{\infty} P[N_a(t) = n] \cdot P[\max(Y_1 \ldots Y_n) > y] .$$

Presuming that a Poisson process gives the number of gusts we have

$$= \sum_{n=0}^{\infty} \frac{e^{-\lambda_a t} (\lambda_a t)^n}{n!} [1 - G_a^n(y)]$$

$$P[X_a(t) > y] = 1 - e^{-\lambda_a t \overline{G}_a(y)} \qquad \text{where } \overline{G}_a = 1 - G_a .$$

Thus we conclude that the time between maximum overloads which exceed a magnitude y, when operating at activity level a, is exponential with occurrence rate $\lambda_a \overline{G}(y)$.

But further if the activity level a(t) is itself a (periodic) function of time, e.g. a(•) maybe the altitude of the ground-air-ground cycle in air transportation, then it can be shown that the maximum overload (gust) that will be encountered in the time interval (o,τ), call it $X^*(\tau)$, has distribution

$$P[X^*(\tau) \leq y] = \exp\{-\int_0^\tau \lambda_{a(x)} \overline{G}_{a(x)}(y)dx\} .$$

There are exact expressions in the case that the activity level follows a regular pattern and further generalization that we do not give.

Model III Cumulative Damage Distributions

Let us now consider the gradual accumulation of damage due to the repetition of duty cycles in each unit of time and for which the incremental damage is itself random. Further we assume that these incremental damages are nonnegative random variables with the same mean μ and variance σ^2; label them

$$X_1, \ldots, X_t, \ldots \qquad t = 1, 2, \ldots, \text{ and then}$$

let $S_t = X_1 + \ldots + X_t$ be the cumulative damage by time t.

Let N be the number of time periods until the cumulative damage exceeds a critical threshold ω and failure occur or untoward action taken. (This would be water accumulating until it runs over a dam.) Then the probability that the time of failure occurs on or before time t is given by

$$P[N \leq t] = P[S_t \geq \omega] \cong \Phi\left(\frac{t\mu-\omega}{\sigma\sqrt{t}}\right) \quad \text{for t large.}$$

The last approximation is made by appeal to the central limit theorem and by physical justification of the approximation. We say that T, a random time of failure due to accumulated damage, has the fatigue life (or the Birnbaum-Saunders) distribution whenever the distribution, call it F, is given by

$$F_T(t) = \Phi\left(\frac{t\mu-\omega}{\sqrt{t}\,\sigma}\right) \equiv \Phi\left[\frac{1}{\alpha}\,\xi(t/\beta)\right] \quad \text{for } t > 0$$

where we write $T \sim FL(\alpha,\beta)$ with

$$\beta = \frac{\omega}{\mu}, \; \alpha = \frac{\sigma}{\sqrt{\mu\,\omega}}, \text{ and } \xi(x) = x^{\frac{1}{2}} - x^{-\frac{1}{2}}.$$

For this distribution one may show that the mean and variance are respectively

$$ET = \beta\left(1 + \frac{\sigma^2}{2}\right), \; Var(T) = (\alpha\beta)^2\left(1 + \frac{5}{4}\alpha^2\right),$$

see [10].

This distribution is rather like the log-normal, except in the tails, since $\xi(x) = -\xi(\frac{1}{x})$ as does $\ln(x) = -\ln(\frac{1}{n})$.

Let us make some comparisons for a random life T of $P[T \geq t]$ for $t > 0$

Log-normal $\ln T \sim N(\ln\beta, \alpha^2)$ $\Phi[\frac{1}{\alpha}(\ln t - \ln\beta)]$

Fatigue Life $T \sim FL(\alpha,\beta)$ $\Phi[\frac{1}{\alpha}\sinh(\frac{\ln t - \ln\beta}{2})]$.

The densities of these two distributions are quite similar, both unimodal and skewed to the right. Moreover the hazard rates of the two distributions are both unimodal but that of the log-normal always approaches zero while the fatigue-life hazard rate converges to $(2\,\alpha^2\beta)^{-1}$.

Last, we consider another cumulative damage model in which damage is assumed to be continuous: at each unit of time the cumulative damage is a normal random variable as is each successive increment and hence capable of assuming negative values with positive probability. Assuming that such momentary repair of damage is possible, perhaps as in concrete, one may postulate that accumulated damage during the interval (o,t) can be described by what is called a Weiner process. Then the time until the damage accumulates to a given threshold has a distribution called the Inverse Gaussian. The distribution (which was first discovered by Schrödinger 1915) has the form

$$G_Y(y) = \Phi[\frac{1}{\alpha}\,\xi(y/\beta)] + e^{2/\alpha^2}\,\Phi[\frac{-2y}{\alpha\beta}\,\xi'(y/\beta)] \quad y > 0 .$$

When parametrized comparably with the fatigue-life distribution we have

$$EY = \beta, \; Var(Y) = (\alpha\beta)^2 \text{ and } Y \sim IG(\alpha,\beta) .$$

Here α is the coefficient of variation and β is the mean life and scale parameter. Surprisingly enough the density $G' = g$ is simple

$$g(y) = \frac{\sqrt{\beta}}{\alpha} y^{-3/2} \varphi[\frac{1}{\alpha} \xi(y/\beta)] \quad y > 0.$$

One notices these comparably parametrized distributions, that $F_T(y) < G_Y(y)$ for all $y > 0$. This means that T is stochastically larger than Y.

But there are statistical advantages to the Inverse-Gaussian distribution in the estimation of its parameters, for like the normal distribution if $Y_1, \ldots, Y_n$ is an i.i.d. sample from $IG(\alpha,\beta)$ then the sample mean is a good estimator of the population mean and moreover

$$\hat{\beta} = \frac{1}{n} \sum_1^n Y_i \sim IG(\frac{\alpha}{\sqrt{n}}, \beta); \; \hat{\alpha} = (\frac{M_1}{M_{-1}} - 1)^{\frac{1}{2}} .$$

Here

$$M_a = (\frac{1}{n} \sum_1^n X_i^a)^{1/a} \quad \text{is the } \underline{ath} \text{ sample mean.}$$

If we set $\theta = \alpha^2/\beta$ then the unbiased minimum variance estimator is

$$\tilde{\theta} = \frac{1}{n-1} \sum_1^n (\frac{1}{Y_i} - \frac{1}{\bar{y}}) = \frac{n}{n-1} (\frac{1}{M_{-1}} - \frac{1}{M_1})$$

and $\tilde{\theta}$ has a known, tabulated statistical distribution. See[9].

If one has an i.i.d. sample from a $FL(\alpha,\beta)$ population then for $\alpha < 1$ (as it is in metallic fatigue problems) it is known that $\tilde{\beta} = \sqrt{M_1 M_{-1}}$, is a very close approximation to the maximum likelihood estimator. This estimate is called the mean mean since it is the geometric mean of the harmonic and arithmetic means. In every case

$$\hat{\alpha} = \left[\frac{1}{n} \sum_1^n \xi^2 (\frac{X_i}{\hat{\beta}}) \right]^{1/2}$$

is the MLE of α. In the case the mean mean can be used to estimate β then

$$\hat{\alpha} = \sqrt{2} \left(\sqrt{\frac{M_1}{M_{-1}}} - 1 \right)^{1/2}$$

See [10].

We now compare the formulae of the densities in these three cases.

$Y \sim IG(\alpha,\beta)$ $\qquad Y \sim G \qquad \xi(x) = x^{1/2} - x^{-1/2}$

$$g(y) = \frac{\sqrt{\beta}}{a} y^{-3/2} \varphi[\frac{1}{\alpha} \xi(y/\beta)] \quad \text{for } y > 0$$

$X \sim FL(\alpha,\beta)$ $\qquad Y \sim F \qquad \xi'(x) = \frac{1}{2}(x^{-1/2} + x^{-3/2})$

$$f(x) = \frac{1}{\alpha\beta} \xi'(\frac{x}{\beta}) \varphi[\frac{1}{\alpha} \xi(x/\beta)] \quad \text{for } x > 0$$

$T \sim LN(\alpha,\beta)$ $\qquad T \sim H$

$$h(t) = \frac{1}{\alpha}\frac{1}{t} \varphi[\frac{1}{\alpha} \ln(t/\beta)] \quad \text{for } t > 0.$$

There are virtually no differences in the regions of central tendency of these distributions. They can all be fitted well to small sets of failure data. But one cannot conclude that these models are interchangeable. Some calculations are easy with one model and difficult with others. Moreover great disparity appears when one calculates the safe-life for a high level of safety using different models. This indicates one must ascertain the basic nature of failure, not just perform curve-fitting to summarize the data.

The construction of any reliability model is based on a clear perception of:

(1) the stochastic nature of the duty cycle (service life) imposed on the component;

(2) the manner in which failure occurs, i.e. either caused by cumulative damage of service loads, the infrequent severe random overload, or both or acts of God;

(3) the conditional probability of failure given the distribution of loads to the component over time, i.e. the right expressions for reliability in the mathematical model;

and (4) the necessity of obtaining loading spectra from components in service which can be used to modify the predictions made by the mathematical models.

As a final illustration let us consider a structural component, selected at random from a population of such, with an initial strength (resistance) S having a distribution G, i.e., $S \sim G$. During the ith time period there is a maximum random load $L_i \sim F_i$ for $i = 1,2,\ldots$ and there is a gradual reduction of the initial strength through cumulative damage processes during each period. Assume that given an initial strength s, the strength during the ith period is the deterministic function $\rho_i(s)$. Hence the probability of not failing during the ith period is

$$P[L_i < \rho_i(s)] = F_i[\rho_i(s)] \ .$$

Thus the reliability for n periods, call it R(n), is given by

$$R(n) = \int_0^\infty \prod_{i=1}^{n} F_i[\rho_i(s)] \, dG(s)$$

assuming that the maximum loads in each period are independent.

An application of this formula is being made to bridge girders where S is the initial (28-day) compressive strength of concrete and ρ_i is the live-load-plus-impact cracking-resistance for i periods as a function of traffic volume; L_i is the maximum load (moment) on the girder during the i time period.

It is generally thought that the closest one may design to the "one-horse shay" is to make strength requirement for all components at the same prescribed level of safety as computed from the system reliability model. Indeed with all components designed to the same level of safety, at a given design life from assumed load distribution, there well may be both savings in weight and increased durability as compared with designing uniformly to the same "safety-factor". After all some persons regard safety-factor and scatter-factor as really measuring ignorance; the higher the one the higher the other.

REFERENCES

1. Handbook of Tables for Probability and Statistics (1976) 2nd Edition, The CRC Press, Inc.

2. Pearson, N. (ed.) (1934) Tables of the Incomplete Gamma Function, Cambridge Univ. Press, Cambridge.

3. Weibull, W. A Statistical Distribution of wide applicability, Journal of Applied Mech. 18 (1951) 293-297.

4. Palmgren, A. Die Lebensdauer von Kugellagern, Z. Ver. Dtsch. Ing. 68 (1924) 339-341.

5. Miner, M. A. Cumulative Damage in Fatigue, J. Appl. Mech. 12 (1945) A159-A164.

6. Freudenthal, A.M. (ed.) Fatigue in Aircraft Structures, (1956) Academic Press Inc., New York.

7. Weibull, W. Fatigue Testing and Analyses of Results, (1961) Pergamon Press, New York.

8. Birnbaum, Z.W., and Saunders, S.C. A Probablistic Interpretation of Miner's Rule, Siam. J. Appl. Math. 16 (1968) 637-652.

9. Chhikara, Raj. S., and Guttman, Irwin, Prediction Limits for the Inverse Gaussian Distribution, Technometrics 24 (1982) 319-324.

10. Birnbaum, Z. W., and Saunders, S.C., A New Family of Life Distributions, Jour. of Appl. Prob. 6 (1969) 319-327.

V. MATHEMATICAL ANALYSIS TECHNIQUES USED IN BUILDING AND CONSTRUCTION TECHNOLOGY

SERVICE LIFE PREDICTION FROM ACCELERATED AGING TEST RESULTS USING RELIABILITY THEORY AND LIFE TESTING ANALYSIS

JONATHAN W. MARTIN

CENTER FOR BUILDING TECHNOLOGY
NATIONAL BUREAU OF STANDARDS
GAITHERSBURG, MD 20899

ABSTRACT

The prediction of the service life of a building product is needed for making life cycle cost decisions, for determining the risks and liabilities in marketing a product, and for helping to identify the causes of a product's failure. Accurate estimates of the service life for building products, however, have traditionally been difficult to obtain. The most timely source for obtaining service life estimates for a building product is through the use of accelerated aging tests. Unfortunately, most accelerated aging tests for building materials are not capable of providing quantitative estimates of a product's service life since they were only designed to screen out bad products or to make qualitative comparisons between the service lives of different products. To rectify this situation, accelerated aging test procedures used in high technology industries were applied to building products. The emphasis in these procedures is in the mathematical analysis of the life data. One of the more successful mathematical procedures is reliability theory and life testing analysis. In this paper, the basic format of this analysis procedure is outlined and several applications are presented in which these techniques were applied to predicting the service lives of different building materials and products. The results of these analyses are very promising and it is concluded that reliability theory and life testing analysis techniques should be readily adaptable for predicting the service lives of a wide range of building products and materials.

1 INTRODUCTION

An accurate estimate of the service life is necessary in order to compute the expected life cycle cost for a building product. Estimates of a product's service life are hard to obtain, however. Standard sources for this information include: 1) a product's performance history, 2) manufacturer's data and claims, 3) long-term outdoor exposure tests of prototype specimens, and 4) accelerated aging tests of prototype specimens.

The performance history of a product has traditionally been the most accurate and authoritative source for service life information. Service life estimates based on this source tend to be fairly accurate, but conservative. They are accurate, because it takes many years of full-scale outdoor exposure testing for a product to establish a performance history. Because of these long exposure times, however, performance history tends to be a very conservative source in that it puts new or innovative products at a competitive disadvantage to products having well-established markets.

If obtaining reliability information from performance history is conservative, then obtaining this information from a manufacturer's claims is the opposite, especially when the manufacturer cannot support his claims with long-term exposure data. For this reason, manufacturer's claims by themselves are seldom the prime source for obtaining estimates of the service life of a product, except in those cases where the reputation of the manufacturer is well-established.

The second most important source of information on the reliability of a building product is the performance of prototype specimens in outdoor exposures. These outdoor exposures are conducted on test racks located in representative, but usually extreme, environments throughout a country. Since the rates of degradation are not greatly accelerated, it normally takes between five and ten years before results are available. This is a long time to wait, and again tends to inhibit innovations and the acceptance of new products.

The last source of information for making service life predictions is accelerated aging tests. Accelerated aging tests are laboratory-based exposures in which one or more of the degradation factors (stress factors) are set to higher levels than normal. The intent, of course, is to increase the rate of material degradation, and thus reduce the cost, extent, and duration of the exposures needed for determining the reliability (hence, the service life) of a product. To date, accelerated aging tests have not played an important role in making service life estimates for building products. This is in contrast to high technology industries, where accelerated aging tests are heavily relied upon. This disparity is due to a fundamental difference in the philosophy of accelerated aging test procedures. These differences are examined further in the next section.

2 ACCELERATED AGING TESTS

The objectives of accelerated aging tests used to evaluate building materials are usually to screen out bad products or to compare a candidate product to products having a long performance history. In high technology industries, the objectives of accelerated aging tests are:

1. to predict quantitatively the service life of a product that is subjected to its intended operating environment. These predictions are made by extrapolating results at high stress levels to expected, in-service levels;

2. to select the best product for a given application based on its predicted service life; and

3. to isolate and identify the causes and mechanisms of early degradation in order to improve the service life of the product.

Implicit in the high technology objectives is an understanding of the interrelationships among: 1) the performance properties of a product, 2) changes in the chemical and physical properties of the materials used in constructing the product, and 3) the stress factors to which the product is exposed. Thus, high technology accelerated aging procedures demand a higher degree of inquiry than those used for building materials. This is better seen when one discusses the two philosophies in terms of the elements of an accelerated aging test.

There are three elements in an accelerated aging test: 1) experiment, 2) evaluation of the degradation state of the specimens, and 3) mathematical analysis of the life data. Each of the elements is vital. In the accelerated aging programs currently used for building materials, the emphasis has been placed on the experimental procedure with comparatively little emphasis being placed on either the evaluation or the mathematical analysis. The apparent philosophy behind this approach has been to develop a laboratory based experimental program that correlates with the results from all or a large portion of outdoor exposures. It is generally agreed that this approach has not been successful.

In the proposed accelerated aging test procedure, more emphasis is placed on the mathematical analysis [1]. Once a mathematical analysis procedure that satisfies the three stated accelerated aging test objectives has been identified, then the requirements for the experimental and evaluation procedures have been stipulated. This emphasis on the mathematical analysis should be expected, inasmuch as the objective of the accelerated aging tests is to quantitatively predict the service life of the building material by extrapolating its performance from high stress levels to those expected in-service. Implicit in these extrapolations, of course, is a very strong role for the materials scientist and chemist in examining and understanding the mechanisms by which specimens fail. If these mechanisms can be identified, then the parameters of the mathematical models, which are used for making the extrapolations from high to low stress levels, can be put in terms of the chemistry and physics of failure, thereby increasing the confidence of extrapolations. Furthermore, it may be possible to eliminate the early causes of failure and thus improve the service life of the building product.

3 RELIABILITY THEORY AND LIFE TESTING ANALYSIS

One mathematical approach that has been extensively used in making service life predictions for a wide range of materials, components, and systems, is reliability theory and life testing analysis. Reliability theory and life testing analysis had its beginnings in the early 1950's. The need for such a theory followed from the results of an audit of the operational readiness of the military's electronic communication systems. The audit showed that the Army and Navy's electronic systems were available less than 30 percent of the time, while the cost of repairing Air Force equipment was over ten times its original cost. Clearly, these results were unacceptable. To help rectify this situation, the military established an ad hoc committee on reliability. From the proceedings of this committee, the scope of the service life prediction problem was better defined and the power of reliability theory and life testing analysis in addressing these problems became better recognized [2].

During the 1950's, the mathematical theory associated with reliability grew rapidly and many successful applications can be cited. The first big success, however, which insured the role of the reliabilist in future projects, was the Minuteman rocket project [3]. In this project, the reliability of large, complex assemblages of electronic components had to be assured for long periods of time. At the start of the program, the mean time between failures for an electronic assemblage containing 10,000 components was 73 h. At the end of the program, the mean time between failures for the same size assemblage had been increased to 20,000 h. This was such a spectacular increase that the reliability techniques used in the Minuteman rocket program were quickly adapted by other research disciplines. Today, reliability theory and life testing analysis techniques are used extensively in the nuclear [4], medical [5], electronic [3], and aerospace [6] industries.

Reliability theory and life testing analysis provide the protocol for both experimentally designing and mathematically analyzing the results of a life test. (A life test is one in which prototypes of a product are subjected to stresses and environmental conditions that typify those to which the product will be exposed in-service. An accelerated life test is a life test in which the intensity of one or more of the stress factors is set at a elevated level. In this paper, the term life test will be used for both the normal and accelerated cases.) The result of this analysis is a prediction of a product's service life. This prediction can be either from normal or from accelerated exposures. Of the two, accelerated exposures are of particular interest to us. Two problems arise in making service life predictions from accelerated aging tests; they are: 1) the large variability in the failure times of individual specimens, and 2) the selection of a stress regime and the extrapolation of the results from a life test at a high stress level to the normal stress level. The first problem is generic to both normal and accelerated aging tests. The second is unique to accelerated aging tests.

In a life test, a number of specimens, say n, are randomly selected from a nominal population and are subjected to a stress S. With increased time of exposure, the specimens begin to degrade. Degradation can result in either a catastrophic failure of a specimen or in a gradual drift out of tolerance of a critical performance property of a product, which is also considered a form of product failure. The time at which a specimen fails is called its failure time. If all of the specimens are placed on test at the same time, then the observed failure times are ordered from the weakest to the strongest; that is, if t_i is the i-th observed time to failure, then:

$$t_1 \leqslant t_2 \leqslant \ldots \leqslant t_i \leqslant \ldots \leqslant t_r \qquad (1)$$

where

r is the number of specimens which have failed at the time the test is terminated, and

n is the number of specimens placed on test, and $r \leqslant n$.

The failure times that are not observed $\{t_{r+1}, t_{r+2}, \ldots, t_n\}$ are termed censored. As is subsequently discussed, censoring plays a very important role in reliability theory.

From the ordered failure times, a sample life distribution can be created (figure 1). In figure 1, the solid dots are the ordered failure times; the tails after the dots are used in evaluating how well a theoretical life distribution fits the sample life distribution. (Strategies for selecting a suitable theoretical life distribution for a material are addressed by Mann et al. [7].)

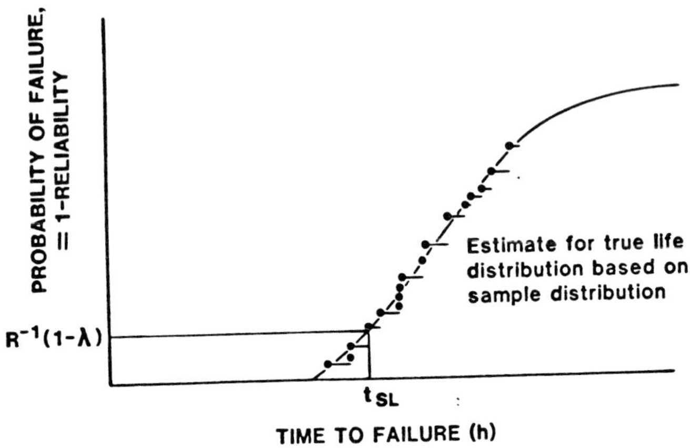

Figure 1. Schematic representation of proposed reliability plan

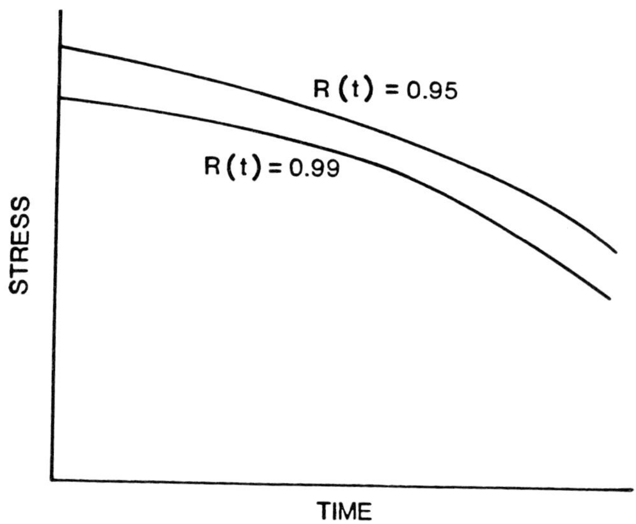

Figure 2. Probability of failure -- stress -- time to failure (P-S-T) diagram. The lines on this diagram are iso-probability lines indicating the reliability or the time beyond which an indicated fraction (.99 and .95) of a nominal population of specimens is expected to survive.

The theoretical life distribution, $F(t)$, is considered to fit the sample life distribution well if it intersects most of the tails following the observed failure times. If a suitable theoretical life distribution can be found, then the vector of parameters, θ, of the theoretical distribution can be estimated from the observed failure times. Since the distributional parameters summarize the information contained in the observed times to failure, and since the number of distributional parameters is usually much lower than the number of observed times to failure, the distributional parameters can be used in place of the observed times to failure, thus greatly simplifying subsequent mathematical analysis.

In accelerated aging tests, the objective is to predict the service life of a product at normal stress levels from its performance at high stress levels. In order to do this, life tests must be conducted over a range of stress levels, S_j, where $j = 0, 1, \ldots,$ m. S_o is the stress level at normal operating conditions, and S_m is the highest stress level at which the mechanism of material degradation is the same as that at S_o. At each stress level, S_j, a set of ordered times to failure is obtained from which the vector of parameters, θ_j, of the theoretical life distribution is estimated. Since the life tests become more severe as the stress level increases (i.e., $S_{j+1} > S_j$, for $j = 1, 2, \ldots,$ m-1), then one would expect:

$$F_{j+1}(t) > F_j(t) \quad \text{for } t>0 \text{ and all } j; \tag{2}$$

that is, for any exposure time, the probability of failure of a product increases as the stress level increases. Assuming equation 2 is valid, it can be shown that a function, $p_j(t;\theta_j)$, exists relating the parameters of the theoretical life distribution to the imposed stress as follows:

$$F_j(t) = F_o(p_j(t;\theta_j)) \tag{3}$$

where

$F_o(t)$ is the life distribution under normal stress conditions. From equation 3, a probability of failure--stress--time to failure (P-S-T) diagram can be constructed (see figure 2). The lines in a P-S-T diagram are iso-probability lines; that is, at a given stress level, each line depicts that time to which a specified proportion of the population is expected to survive. As the stress level decreases, this time increases; thus, the P-S-T diagram gives estimates of the service life of a product over a wide stress range. Hence, the P-S-T diagram can be used by a consumer in determining the expected service life for a product at the stress level corresponding to a given geographical location, or in making selections between competing products.

Presenting the times to failure of the specimens in the form of a distribution raises such issues as which time to failure should be designated as the service life for a product; whether certain times to failure should be assigned more weight in fixing a product's service life; whether given the relative importance of the different times to failure, a strategy for improving the service life of a product is suggested; and what is the effect of censored times to failure in predicting the service life for a product.

Selecting the time to failure to be designated as the service life for a product is not a trivial problem -- especially if, as is often the case, the range of failure times covers several decades of time. Of the products sold, the manufacturer is most liable for those that fail prematurely. In order to minimize his liability, a manufacturer may select that time as the service life at which only a small percentage of the population of a product is expected to fail. The percentage chosen would be a function of such factors as safety, accessibility, warranty costs, and importance of the product. For example, in the case of a coating system, the selected percentage may be very small for a coating system applied to the towers of a suspension bridge; whereas, a much larger percent may be selected when a coating system is to protect a small building.

Since the service life of a product is likely to be defined as that time at which some small percentage of the population is expected to fail when subjected to its expected operating conditions, it is obvious that the early failure times are given more weight in determining the service life of a product than the longer failure times. In order to improve a product's service life, therefore, one should attempt to identify all causes of the early failure times. If these mechanisms can be identified and eliminated, then the product's service life should increase.

In most life tests, the experiment will have to be terminated prior to observing the failure of all of the specimens; hence, some specimens will have censored failure times. The degree of censoring will depend on many factors including the stress level, the number of specimens on test, the duration of the life test experiment, and changes in the failure rate with product age. For a given sample size, confidence in the estimates of the life distribution parameters, θ, decreases as the degree of censoring increases. The confidence one has in the estimate of a parameter depends, however, both on the number of specimens tested, n, and on the number of censored times to failure. Hence, it is possible to have the same confidence in a parameter by either having a large number of specimens on test with a large number of censored specimens or by having a small number of specimens on test with a few censored specimens. The advantage in having a large number of specimens on test with a large number of censored specimens is that the life test can usually be terminated in a shorter time. Therefore, if the cost of making specimens is low and the cost of labor or of delay high, the best strategy is often to use a large number of specimens [8].

4 EXAMPLES OF THE APPLICATION OF RELIABILITY THEORY TO BUILDING MATERIALS

The number of successful applications of reliability theory and life testing analysis techniques to all materials in all disciplines is quite large. Many of these examples are cited by Mann et al. [7]. For building materials, this technology is new, so few examples are currently available. This number is increasing, however, as these techniques become adopted by the building research community. In this survey only the author's research or research on building materials conducted at the National Bureau of Standards is cited. A more extensive review for polymeric materials is given by Martin [9].

4.1 Service Life Prediction for Coatings on Steel Substrates

Accelerated aging tests have not played an important role in predicting the service life of coating systems on steel [10]. The reason for this is, in part, that there has been little need for accelerated aging tests, or even for new coating systems, prior to the 1970's, since the industry had two systems (red lead and basic lead silico-chromate oil alkyds) that were prescribed for almost all applications because of their good performance histories.

In the late 1960's and 1970's, however, a number of federal and state regulations were established in the United States that restricted the concentrations of solvents and the levels of certain pigments in coatings. Since the two traditional coating systems were affected on both counts, manufacturers immediately began to formulate and market new, high-performance coating systems. Since these systems were new, no estimates of their expected service lives were available, so consumers were reluctant to purchase them. An obvious need existed, therefore, for obtaining accurate service life predictions for these systems prior to the implementation of the restrictions on solvent concentration and chemicals.

Since these high performance coating systems had limited performance histories, extensive outdoor exposure tests were started. Outdoor exposure tests, however, are expensive to conduct and take a long time before results are available. Since the number of material, processing, and exposure variables is very large, it was almost certain that little service life data would be available for the new coating systems by the time the new regulations were implemented, unless valid accelerated aging test methodologies were developed. One way of improving the accelerated aging tests for coating systems was to adopt the proven techniques of reliability theory and life testing analysis.

In order to apply these proposed reliability techniques to coatings, several radical changes had to be made to the current accelerated aging test procedures [11]. The major change was to replace the visual evaluation technique for measuring the proportion of defects with one having a stronger statistical basis. The new evaluation procedure is based in spatial statistics [12], is more quantitative and systematic, and has higher resolving power than the current visual procedures [11,13,14]. This evaluation procedure appears to be both repeatable and reproducible, so that the output from this analysis can be used directly in the reliability analysis.

Two coating systems -- an alkyd and an acrylic -- were used to verify the new procedures. These coatings were applied to hot-rolled steel panels having a profile of 5 mils (.002 cm). Forty-five panels from each coating system were divided among three temperatures and exposed at 95 percent relative humidity (2 coating systems X 3 temperatures X 1 relative humidity X 15 panels = 90 panels). For each panel, the percent of corroded, blistered, and corroded plus blistered areas were measured and tracked as a function of exposure time. A panel was designated as having failed when 0.3 percent of its area had degraded; this is equivalent to an ASTM visual corrosion grade of 7 [13]. The failure time for a panel, therefore, is the time at which the panel first exceeds this limit. Corrosion plus blister tracks for each temperature are shown in figure 3 for the alkyd coating system.

From figure 3, the sample life distributions for corrosion plus blisters with the alkyd coating at each temperature were determined and plotted (figure 4). To each sample life distribution, a Weibull distribution was fitted. The Weibull distribution has two parameters, a scale parameter, beta, and a shape parameter, alpha. These two parameters act like the mean and the variance of the normal distribution in that they affect the spread and the shape of the distribution. In figure 5a, the logarithm of the scale parameter has been plotted against reciprocal temperature, using an Arrhenius model. The shape parameter was plotted against temperature in figure 5b, using a power law. These same equations with different coefficients were used in modeling the effects of temperature on the acrylic coating system. From these equations, a P-S-T diagram for the alkyd coating system was constructed, figure 6. The P-S-T diagram can be used by

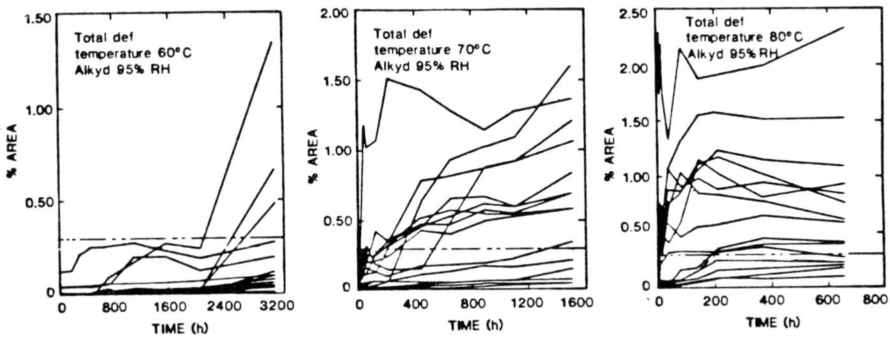

Figure 3. Proportion of area that contains corrosion plus blisters as a function of time for alkyd coating system exposed to 95 percent relative humidity and one of three temperatures (60, 70 and 80 C)

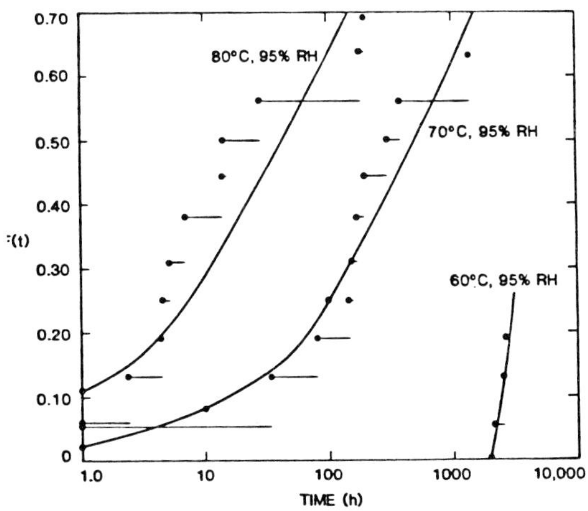

Figure 4. Sample amd theoretical life distributions for alkyd coating system exposed to 95 percent relative humidity and one of three temperatures (60,70,and 80 C)

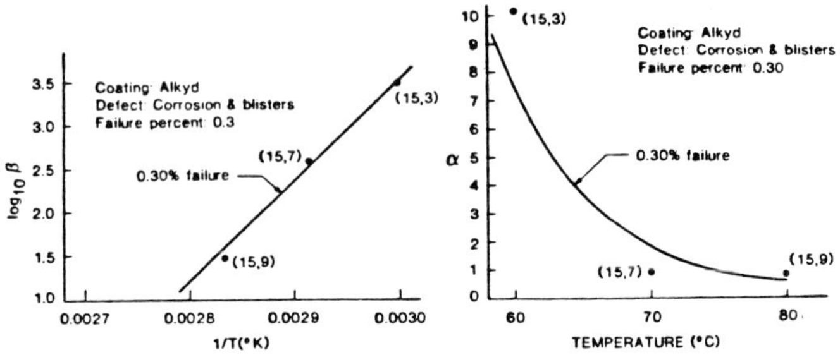

Figure 5. Regression line fit to data for alkyd coating system using .3 percent area corrosion plus blisters as the failure criterion; (a) common logarithm of Weibull scale parameter versus reciprocal absolute temperature, and (b) Weibull shape parameter versus temperature.

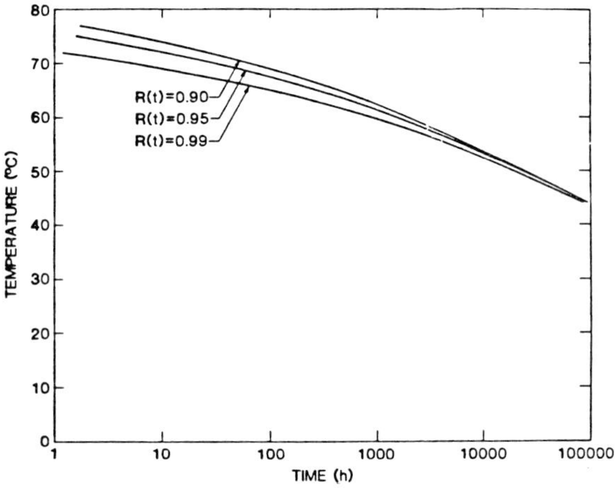

Figure 6. Probability of failure--temperature--time to failure diagram for alkyd coating system, using .3 percent area corrosion plus blisters as the failure criterion

both consumers and manufacturers of coating systems to determine the service life of a coating system over a wide range of temperature and at any desired level of reliability. A more complete description of the analysis is given by Martin and McKnight [15].

4.2 Photolytic and Thermal Degradation of PMMA Films

Poly (methyl methacrylate) or PMMA is a linear polymer that has been extensively studied with respect to its chemical and physical properties over the last thirty years. In practice, PMMA is used for such applications as signs, windows, and automobile tail light reflectors. PMMA was also a candidate material for cover plates in active solar collectors [16]. This study was initially undertaken to determine the applicability of PMMA as a cover plate material. As the research progressed, however, this objective became of secondary importance, and the new objective of the research became to utilize PMMA as a model material for demonstrating the applicability of reliability theory and for understanding the synergistic effects of several stress factors on the service life of PMMA sheets. The stress factors studied were temperature and ultraviolet radiation.

Experimentally, pieces of commercial and pure PMMA sheets were exposed to one of six temperatures (50, 70, 85, 100, 115, and 125 C) and one of four irradiance levels (24 cm, 30 cm, and 37 cm from the source, and no radiation). The primary time-dependent measures of degradation were changes in the number average molecular weight, M_n, and changes in the ultraviolet spectral transmittance. The number average molecular weight is strongly correlated with a film's glass transition temperature, and with strength and other physical properties. In the case of a cover plate material, the film tends to embrittle and creep with decreasing average molecular weight. Failure of a specimen is defined as having occurred when the number average molecular weight, M_n, of a specimen falls below a minimum threshold value (see figure 7).

For a population of PMMA films, a variation in the initial number average molecular weights is often observed. This initial variation results from measurement errors and, for polymer sheets, variations introduced during fabrication. During exposure of the film, other sources of variation are introduced that result from differences in the degradation rates of different specimens. These variations are depicted as probability density functions in figure 7. With continued exposure, the density functions are shifted toward the threshold value and thus, some specimens from the population begin to fail. From reliability theory, the service life for the nominal population of PMMA films is defined as that time at which an unacceptable proportion, $1 - \bar{H}(t)$, of the population fails, where $\bar{H}(t)$ is the survival fraction.

A model for the survival fraction, $\bar{H}(t)$, was developed for films subjected to both thermal and photolytic degradation. The photolytic degradation of PMMA was modeled as a Poisson process [17]. Conceptually, reductions in the average molecular weight result from photon induced random chain scissions. The Poisson distribution models the probability that exactly K photon-related chain scissions will occur in the interval [0,t). Thermal degradation of the film was modeled as a weak-link process. The number of thermally caused random chain scissions was assumed to be normally distributed. Using the normal approximation to the Poisson distribution, the probability of surviving the combined effects of thermal and photolytic degradation for a time t was shown to be [18]:

$$H(t) = \int_{-\infty}^{\infty} F_k(M_n') \left(\frac{1}{2\pi(\sigma^2 + \lambda t)}\right)^{1/2} \exp - \left[\frac{(k-\beta_T - \alpha_T t-\lambda t)^2}{2(\lambda t + \sigma^2)}\right] dk. \quad (4)$$

where

$F_k(M_n') = \int_{M_n'}^{\infty} f_k(\mu)d\mu$ and is the probability that the average molecular weight has a value greater than the minimum allowable number average molecular weight and M_n' is the minimum acceptable number average molecule weight;

$$\lambda = \left[\exp\left[-\left(\frac{E(T_o - T)}{RT_o T}\right)\right]\right]\left[\left(\frac{I_a}{I_\alpha}\right)^n\right]\Delta_o \quad \text{and}$$

is the Poisson parameter for the expected number of chain scissions per unit time resulting from photolytic degradation;

and

α_T and β_T are coefficients of a weak link equation for modeling the number of random chain scissions resulting from thermal degradation.

The Poisson model for photolytic degradation was validated against 25 published data sets. The model appears to be consistent with these published data and the results of other researchers [17]. However, the combined model has not yet been fully validated because of some unexpected experimental results including the occurrence of thermal degradation, which had not been previously reported at these low temperatures. Research is continuing in order to isolate the causes of this thermal degradation and to validate the full model.

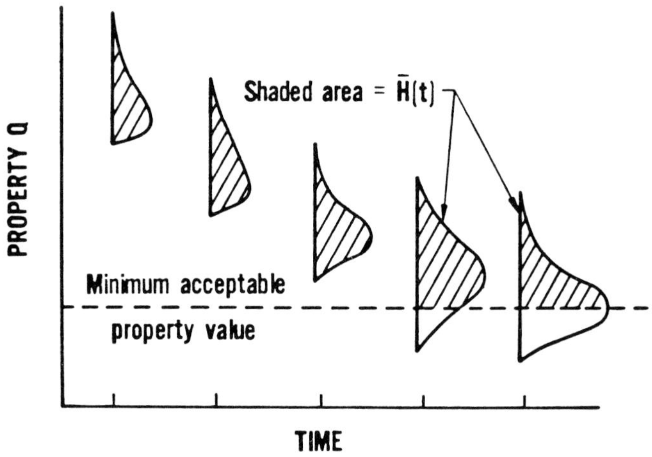

Figure 7. Schematic showing changes in the density function of a critical performance property of a building product with duration of exposure to its intended operating and service environment.

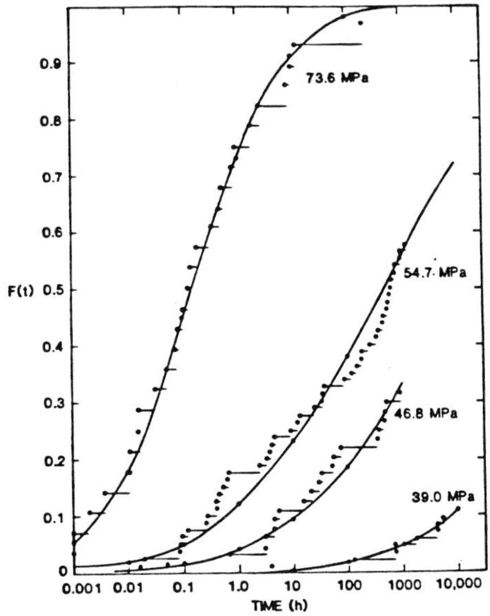

Figure 8. Sample and theoretical life distributions for small clear wood specimens exposed to one of four stress levels (39.0,46.8,54.7, and 73.6 MPa)

4.3 Long-Term Heat Storage Efficiency of Phase Change Materials for Solar Energy Applications

Phase change materials have a potential application for space heating in buildings, as long as the equilibrium transition temperature of the material falls within the diurnal temperature range. When the temperature of the material passes through its equilibrium transition temperature, it either stores energy (on heating) or releases energy (on cooling) in the form of latent heat. One class of these phase change materials is hydrated salts, and an example is Glauber's salt, $Na_2SO_4.10H_2O$.

Ideally, with each thermal cycle, the phase change should be 100 percent efficient, i.e, return to its hydrated form. In practice, however, these phase transitions are not that efficient; instead, after a number of cycles, segregation is observed, which reduces the freezing temperature to 11 C, at or below the useful temperature range for space heating. To prevent segregation, additives are mixed with the phase change material. These additives include thickening agents to minimize the settling of Na_2SO_4 during cooling, nucleating agents to facilitate the hydration of Na_2SO_4 during cooling, and crystal habit modifiers to control the size and surface area of the Na_2SO_4 crystals. Even with these additives, however, the phase change system still segregates after a number of thermal cycles. The result of segregation is a large decrease in the thermal storage capacity of the phase change material, making the system uneconomical for space heating applications. The objective of this research was to investigate the reasons for this segregation.

The research was divided into two parts. The first part sought to understand the material aspects of the problem [19], while the second part was aimed at stochastically modeling the changes in mass fraction of each constituent of the phase change system at the end of each cycle [20]. The stochastic model is completed and is currently being tested against the experimental data.

4.4 Duration of Load for Wood Structural Members

In the United States, as well as in Europe and Canada, a gradual move is underway to adopt Limit States Design Concepts in the design of structures [21]. One of the limit states design formats is called Load and Resistance Factor Design. This format incorporates into its methodology limits states for resistance and load factors. With wood, the resistance factor for duration-of-load is one of the more significant factors in determining the capacity of a wood structural element. It is a difficult factor to determine though for structurally-sized members.

The duration-of-load factor currently used for wood is based on the results of experiments conducted on small clear specimens [22]. With the increased pressure to implement Limit States Design Concepts for all building materials, the question arose as to whether structural wood members have the same duration-of-load factor as small clear specimens.

To test this, it was necessary to conduct duration-of-load tests on structural members. Conceptually, this was easy; experimentally several difficulties arise. The major problem was that the experimental procedure used for small clear specimens did not work for structural specimens. For small clear specimens, the experimental procedure relied on a matched specimen for determining the ultimate strength of the duration-of-load specimen. The duration of load specimen was then loaded to a predesignated fraction of its (actually its matched pairs) ultimate strength. For structural members, a matched specimen does not exist; therefore, a new experimental and analytical procedure had to be developed. Martin and Saunders [23] developed a procedure based on reliability theory. The procedure was shown to be mathematically equivalent to the one used for small clear specimens.

To validate this procedure, an experiment was run using a nominal population of small clear specimens. Specimens were randomly selected from this population and assigned to one of four stress levels (39.0, 46.8, 54.7, and 73.6 MPa). The failure time for each specimen was observed, recorded, and the sample life distribution constructed at each stress as shown in figure 8. A lognormal distribution was fit to each sample life distribution. The two parameters of the lognormal distribution were plotted versus mechanical stress (figure 9). The regression equations used in fitting the parameter versus stress diagrams were then used in computing a P-S-T diagram (figure 10). As expected, the results of this experiment were consistent with those using a pair-matching procedure [24]. More importantly, some preliminary results indicate that the proposed procedure is also valid for structurally-sized members [25].

5 RESEARCH IN PROGRESS

Reliability theory and life testing analysis techniques are being applied to other building materials. Two on-going projects are the delamination of adhesively bonded honeycomb sandwich panels and the debonding of single-ply roofing material joints. These programs are in their early experimental stages, and the results are not yet available.

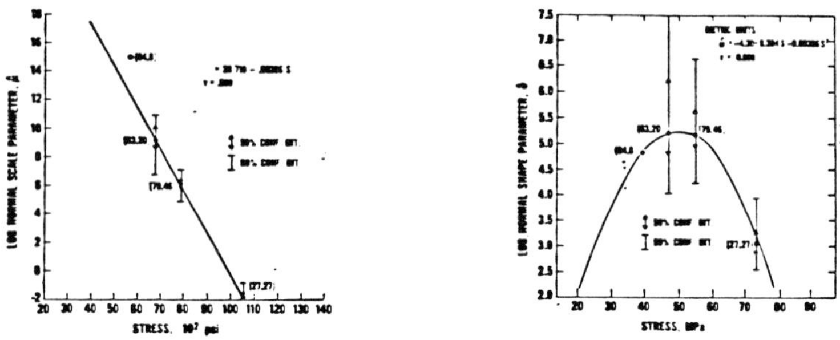

Figure 9. Regression line fit to data for small clear wood specimens; (a) lognormal scale parameter versus mechanical stress, and (b) lognormal shape parameter versus mechanical stress.

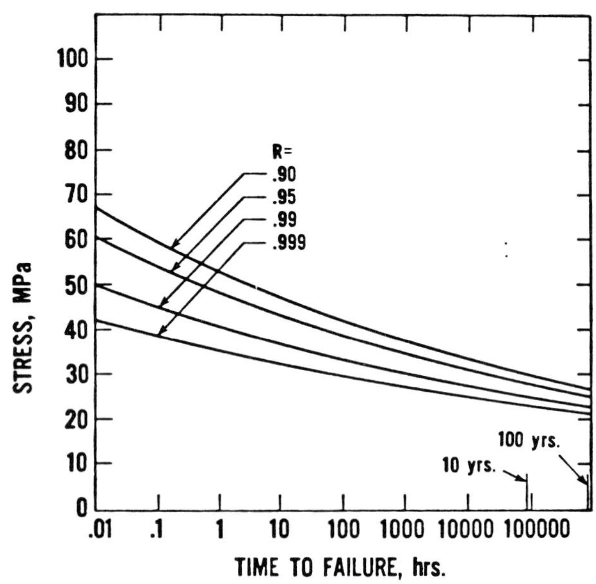

Figure 10. Probability of failure --mechanical stress-- time to failure diagram for small clear wood specimens.

6 CONCLUSIONS

Reliability theory and life testing analysis techniques have been applied to predicting the service lives for several building materials from the results of accelerated aging tests. The results of these analyses are very promising,and it is concluded that these techniques should be readily adaptable for predicting the service lives of a wide range of building products and materials.

7 REFERENCES

1. ASTM E632, Standard Practice for Developing Accelerated Tests to Aid in the Prediction of the Service Life of Building Components and Materials, American Society for Testing and Materials, 1982.

2. O'Connor, P. D. T., Practical Reliability Engineering, Heyden, Philadelphia, 1981.

3. Shooman, M. L., Probabilistic Reliability: An Engineering Approach, McGraw Hill, New York, 1968.

4. Vesely, W. E.; Goldberg, F. F.; Roberts, N. H.; Haasl, D. F., Fault Tree Handbook, U.S. Nuclear Regulatory Commission, NUREG-0492, 1981.

5. Miller, R. G., Survival Analysis, Wiley, 1981.

6. Whittaker, I. C.; Besumer, P. M., A Reliability Analysis Approach to Fatigue Life Variability of Aircraft Structures, Air Force Materials Laboratory Tech. Rep. AFML-TR-69-65, 1969.

7. Mann, N. R.; Schafer, R. E.; Singpurwalla, N. D., Methods for Statistical Analysis of Reliability and Life Data, Wiley, New York, 1974.

8. Epstein, B.; Sobel, M., Life Testing, J. Am. Stat. Assoc., 48: 486-502, 1953.

9. Martin, J. W., Time Transformation Functions Commonly Used in Life Testing Analysis. Durability of Building Materials 1: 175-194, 1982.

10. Campbell, P. G.; Martin, J. W.; McKnight, M. E., Short-term Evaluation Procedures for Coatings on Structural Steel, National Bureau of Standards, Tech. Note 1149, 1981.

11. Martin, J. W.; McKnight, M. E., The Prediction of the Service Life of Coatings on Steel, Part I: Procedure for Quantitative Evaluation of Coating Defects, Paper submitted to J. Coatings Tech.

12. Ripley, B. D., Spatial Statistics, Wiley, New York, 1981.

13. ASTM D610, Standard Method of Evaluating Degree of Rusting on Painted Steel Surfaces, American Society for Testing and Materials, Part 27, 1983.

14. ASTM D714, Standard Method for Evaluating the Degree of Blistering of Paints, American Society for Testing and Materials, Part 27, 1983.

15. Martin, J. W.; McKnight, M. E., The Prediction of the Service Life of Coatings on Steel, Part II: Quantitative Prediction of the Service Life of a Coating System, Paper submitted to the J. Coatings Tech.

16. Clark, E. J.; Roberts, W. E.; Grimes, J. W.; Embree, E. J., Solar Energy Systems--Standards for Cover Plates for Flat Plate Solar Collectors, National Bureau of Standards, NBS Tech. Note 1132, 1980.

17. Martin, J. W., A Stochastic Model for Predicting the Service Life of Photolytically Degraded Poly (Methyl Methacrylate) Films, J. Appl. Poly. Sci. 29(3): 777-794, 1984.

18. Martin, J. W.; Waksman, D.; Bentz, D. P.; Lechner, J. A.; Dickens, B., A Preliminary Stochastic Model for Service Life Prediction of a Photolytically and Thermally Degraded Polymeric Cover Plate Material, Proc. Third International Conference on the Durability of Building Materials and Components, Espoo, Finland, 1984.

19. Brown, P. W.; Grimes, J. W.; Kaetzel, L., Evaluation of the Variation in Thermal Performance in a $Na_2SO_4 \cdot 10H_2O$ Phase Change System, Paper to be submitted to J. Solar Energy Materials.

20. Saunders, S. C., A Mathematical Model for the Distribution of Long-term Heat-storage Efficiency of Phase Change Materials, Internal Document of Scientific Consulting Services, Pullman, Wash., 1984.

21. Ellingwood, B.; Galambos, T. V.; MacGregor, J. G.; Cornell, C. A., Development of a Probability Based Load Criterion for American National Standard A58, National Bureau of Standards, NBS Special Publication 577, 1980.

22. Wood, L. W., Relation of Strength of Wood to Duration of Load, USDA Forest Service Paper No. 1916, 1951.

23. Martin, J. W.; Saunders, S. C., The Relationship Between Life Test Analysis and Tests at Constant Stress Ratios, in Structural Use of Wood in Adverse Environments, Eds., R.W. Meyer and R.M. Kellogg, Van Nostrand, New York, 1982.

24. Martin, J. W., The Analysis of Life Data for Wood in the Bending Mode, Wood Science and Technology 14: 187-206, 1980.

25. Unpublished analysis of Barrett and Foschi's duration of load data on structural-sized wood specimens.

MATHEMATICAL MODELLING OF MICROSTRUCTURE AND PROPERTIES OF HYDRATED CEMENT

Leslie J. Parrott

Cement and Concrete Association, U.K.

ABSTRACT

Changes in building methods and materials make it increasingly difficult to devise simple and reliable procedures for the design of projected structures that reflect the engineering experience gained from existing and failed structures. This is particularly true for the durability of concrete structures.

The durability of concrete structures is normally controlled by diffusion of harmful fluids through an exposed concrete surface. A model is described which allows the microstructure and moisture state of the exposed surface region to be quantified. The model can thus be used to illustrate how engineering parameters such as mix proportions, curing time and reinforcement cover can affect the physical aspects of durability.

Preliminary comparisons between the model and experiment show that moisture content, cement hydration and porosity predictions are encouragingly realistic. However it is emphasised that further verification of the model is required and that the model should be developed to include carbonation.

1 INTRODUCTION

The construction industry has relied traditionally upon engineering experience to ensure that structures will remain serviceable throughout their design life. However the design life of a structure may be long in relation to the shortening time scales associated with current changes in building methods and materials.

Thus it is becoming increasingly difficult to ensure that engineering experience is adequate and to formulate design code rules that are both reliable and simple. One approach to overcome this difficulty is to develop models for the prediction of certain aspects of serviceability that are broad enough in scope to encompass the performance of old and of projected structures. This would allow information from existing and failed structures to be used to establish the credibility of the predictive models. It would also encourage the application of basic research findings to the formulation of the predictive equations.

Reliable design methods have been developed that ensure the ultimate load-bearing capacity of a structure is adequate and that structural movements can be predicted with sufficient accuracy for most purposes. The durability of concrete structures is less well understood and, at the present time, it is difficult to quantify. Laboratory tests, often accelerated, may allow ranking of materials with regard to some particular aspects of durability but they will rarely simulate the conditions under which a projected structure will be exposed. This deficiency has led to a number of field studies in recent years where exposure conditions are obviously realistic but the range of engineering variables (such as curing period, mix proportions and cover over the reinforcing steel) is limited. It would be helpful if laboratory and field data could be synthesised to provide a method to predict the progressive loss of serviceability. Design parameters could then be selected such that particular, deleterious processes do not reach a critical stage during the design life of a structure.

2 SCOPE

The major part of this report will describe a model of the time-dependent changes in moisture state, hydration and porosity that develop in concrete when it is exposed to a drying environment. A clear picture of the variations in these parameters with distance from the exposed face is regarded as a prerequisite for engineering formulations in which the development of various, deleterious processes are to be considered. A central consideration of the model is the moisture balance in the cement paste phase of each spacial element, under conditions of uniaxial drying and constant temperature. Outputs are the evaporable water content, chemically bound water content, relative humidity and porosity for each element after a succession of time increments. Currently the input parameters are exposure relative humidity, length of drying path, initial curing time, water/cement ratio, aggregate/cement ratio, aggregate absorption and fraction of pulverised fuel ash (pfa) blended in the cement. Many other desirable input parameters have not yet been considered: it seemed preferable to demonstrate the utility of a relatively simple model before embarking upon a more sophisticated study.

Development of the model includes two areas of work that are not covered in detail in this report, but are believed to be essential for the credibility of any predictions. First, experimental studies are in progress wherein all of the main output variables are being measured for selected sets of input parameters. These measurements are being compared with predictions as a check on the reliability of the model and to locate any inadequacies. Second, basic studies have been initiated to provide more reliable relationships between the variables used in the modelling. For example the functional relationships between rate of cement hydration and relative humidity, cement composition and degree of hydration are being examined.

Predictions will be compared with two sets of experimental data to show that the model, even in its present unrefined state, yields reasonable values of evaporable water content, bound water content and porosity.

Finally the following topics will briefly be considered

a) Effect upon microstructure of the surface regions of replacing cement with pfa.

b) Carbonation and its effect upon porosity gradients.

c) Design for durability.

3 DESCRIPTION OF MODEL

3.1 Inputs

The input parameters that describe the concrete are the water/cement ratio (w/c), aggregate/cement ratio, aggregate absorption and the fraction of pulverised fuel ash blended with the Portland cement (pfa). The concrete element is defined in terms of the length of its uniaxial drying path. Exposure is considered in two stages; the first stage assumes sealed storage for the selected curing time. The second stage involves drying at a selected relative humidity (R).

3.2 Hydration

Cement hydration is considered here in terms of the weight of water that becomes chemically bound with unit weight of cement, WN. This view of hydration is particularly convenient when considering changes in the moisture content of an element. The bound water in an element, WN_T, is obtained by integration, over a time increment ΔT, of the equation

$$WN_T = WN_{T-1} + \Delta T . \dot{WN} \qquad 1$$

where $\dot{WN}$, the rate of bound water development, is calculated from the last value of bound water WN_{T-1} and the relative humidity, R_{T-1} according to the equation

$$\dot{WN} = 10.37\,(WNM - WN_{T-1})^3 \left(\frac{R_{T-1} - 0.55}{0.45}\right)^4 \qquad 2$$

The first part of equation 2 represents the rate of bound water development under saturated conditions quite adequately if ΔT is maintained at 0.5 days for the first day or two of curing. The second part of equation 2 describes the reduction in the rate of hydration when the relative humidity is less than saturated and is based upon reference 1. The maximum value of bound water, WNM is obtained from

$$WNM = 0.24\,(1 - 0.375\ pfa) \qquad 3$$

It is fortuitous that the effect of pulverised fuel ash upon bound water development can be described by merely altering the value of WNM in equation 2. Figure 1 illustrates experimental values of bound water development under saturated conditions together with the theoretical curves based upon equation 1.

A more sophisticated representation of hydration has recently been devised in which the effects of water/cement ratio, cement composition and specific surface area are additionally accounted for (1). This is currently being incorporated into the model.

3.3 Porosity

The porosity of the cement paste is subdivided into large and small pores as indicated in Figure 2. The small pores increase in volume with an increase in bound water content and are associated with the gelatinous products of cement hydration. The large pores originate from the space left by the mix water and the space vacated by the hydrated cement. These spaces are partly filled with the products of hydration and their associated small pores. This description of the phases in hydrated cement derives from the work of Powers and Brownyard (2). More recent studies suggest that the porous gel in hydrated cement will shrink, even with quite moderate drying (3). The shrinkage of the gel seems to derive from a reduction in volume of the small pores, without any change in total porosity. Thus the volume of large pores is increased and the rate of diffusion is correspondingly increased (4). In the present study, where we are concerned with moisture diffusion, the emphasis will be on the relative volume of large pores, P: it is conveniently calculated from the difference between the total porosity, P_1 and the small porosity, P_2. The small porosity is defined here as the volume fraction of pores less than 4 nm wide.

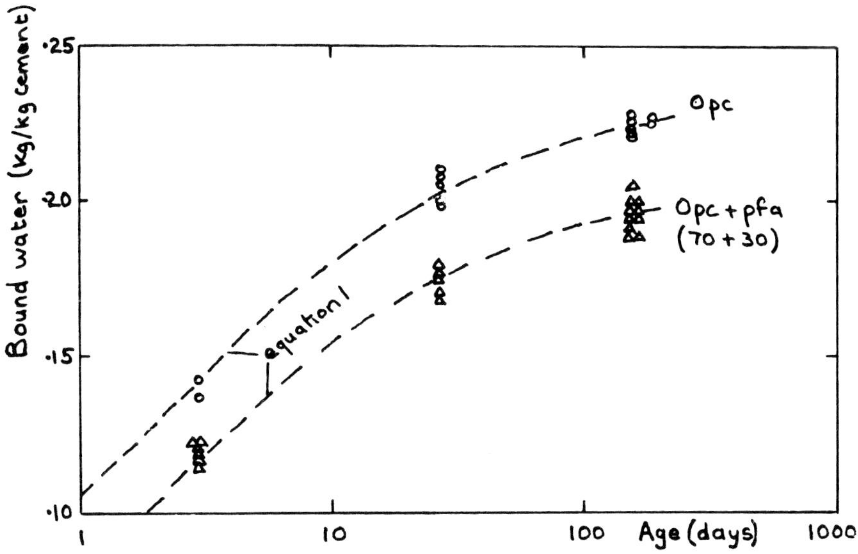

FIGURE 1 EXPERIMENTAL AND THEORETICAL DEVELOPMENT OF BOUND WATER

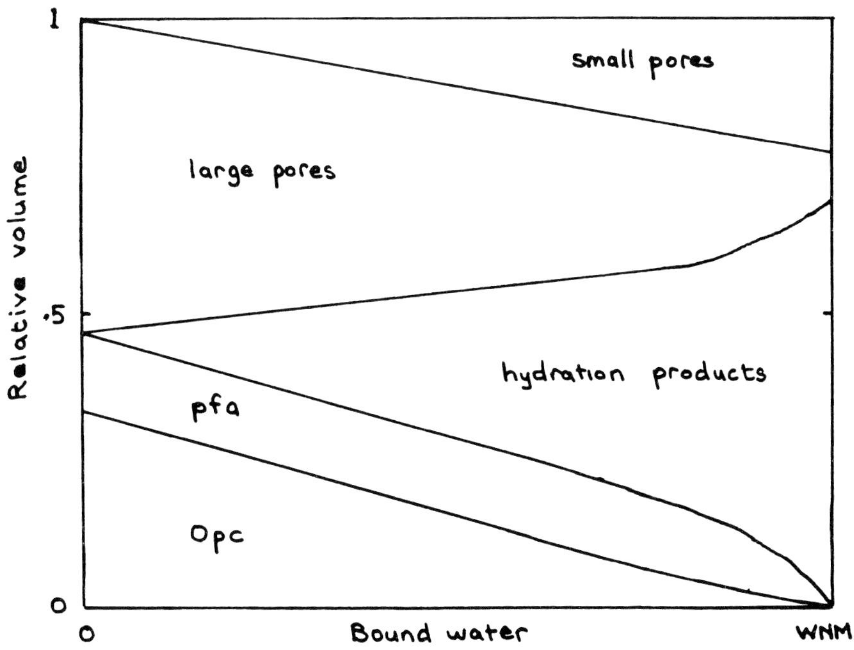

FIGURE 2 DIAGRAM OF PHASES IN HYDRATED CEMENT PASTE

$$P_1 = 0.001.C\,(W/C - WN\,(0.644 - pfa)) \qquad 4$$

$$P_2 = 0.001.C.f(R).\left(1.061\,WN + \frac{pfa}{1.1}\left(\frac{WN}{0.244}\right)^8\right) \qquad 5$$

C is the cement content in kg/m^3 of paste and f(R) is a function that represents the effect of drying upon the volume of small pores. f(R) is based upon recent experimental work (3,4) and for values of relative humidity down to 0.5 it can be evaluated with equation 6

$$f(R) = 0.7 + 1.2\,(R - 0.5)^2 \qquad 6$$

The numerical constants in equations 4 and 5 are based upon unpublished experimental data for ordinary Portland and ordinary Portland/pfa blended cements.

3.4 Evaporable water

The change of evaporable water content, ΔEW in an element is calculated for each time step, ΔT using the following relationships

$$\Delta EW = \Delta T\,(Q_2 - Q_1 - Q_3) \qquad 7$$

where $$Q_1 = K_1 \frac{\Delta R1}{\Delta X}, \quad Q_2 = K_2 \frac{\Delta R2}{\Delta X}$$

and $$Q_3 = C.\,\Delta WN$$

Q_1 and Q_2 represent the uniaxial flow rates of evaporable water out of and into the element by virtue of the relative humidity gradients $\frac{\Delta R1}{\Delta X}$ and $\frac{\Delta R2}{\Delta X}$. Q_3 represents the reduction of evaporable water due to the increase in bound water content, ΔWN. K_1 and K_2 are diffusion coefficients.

The flow rates of evaporable water could have been estimated from gradients of evaporable water content rather than gradients of relative humidity. However this approach was rejected because it would lead to a final "equilibrium" condition where the uniform level of evaporable water content would be associated with a non-uniform relative humidity profile. This arises from the non-uniform porosity profile that is implicit in the model.

The diffusion coefficients are critical parameters in the performance of the model but they are difficult to evaluate from experimental measurements because the porosity of hydrated cement changes during drying. A further complication is that the coefficient appears to be very sensitive to small changes in relative humidity at conditions near to saturation and experimental control under such conditions is difficult. Published literature on water diffusion (5,6,7) and some preliminary comparisons between predictions and experiment suggested that the following relationship might be applicable

$$\text{Diffusion coefficient, } K = 2500 . P . \left(\frac{R}{1000} + \left(\frac{R-0.4}{0.6}\right)^{14}\right) \quad 8$$

This equation implies that moisture transfer is very rapid where conditions are close to saturation and reflects, in a simple way, the influence of porosity. It should be emphasised that much more experimental work is required to obtain a better representation of the diffusion coefficient.

3.5 Relative Humidity

The calculated changes of evaporable water content imply corresponding changes in relative humidity and indeed new values must be provided before the set of calculations for the next time step can be performed. As with porosity the relationship between evaporable water content and relative humidity is based upon a modification of the work of Powers and Brownyard (2). Two components of evaporable water content are calculated from the equations

in small pores

$$EW1 = C . f(R) . 0.32 . \frac{WN}{WNM} . R^{0.4} \quad 9$$

in large pores

$$EW2 = C . \left(W/C - \frac{WN}{WNM}\left(0.5 - 0.32\left(1 - f(R) . R^{0.4}\right)\right)\right) . \left(\frac{R-0.4}{0.6}\right)^{4} \quad 10$$

These equations are similar to those used in the porosity calculations but include the functions $R^{0.4}$ and $\left(\frac{R - 0.4}{0.6}\right)^{4}$ to represent the extent to which the small and large pores are water-filled. Equations 9 and 10 are used to provide successive estimates of evaporable water content, EW1 + EW2, with trial values of R until the estimate matches the required figure. A binary selection procedure is used to obtain trial values of R, thus giving rapid convergence between the estimated and required values of evaporable water content.

3.6 Aggregate

In saturated, freshly mixed concrete a significant volume of water can be held in the pores of the aggregate and may be removed later by external drying or by the self-desiccating action of cement hydration. Aggregate/cement ratio and aggregate absorption are included as inputs so that allowance can be made for this additional volume of water. The model assumes that the aggregate is initially saturated and thereby provides an additional volume of evaporable water. Furthermore it is assumed that aggregate porosity does not significantly increase diffusion coefficients.

3.7 Calculation Sequence

The input variables are entered and the bound water content development is then calculated for the duration of the curing period, assuming no loss or gain of water. Common initial values of bound water, evaporable water, porosity and diffusion coefficient are then calculated for each spacial element. The drying path length (i.e. half the total member thickness) is normally divided into ten elements of equal thickness, DX.

The iterative part of the calculation involves the following steps:

a) Calculate time increment. This is gradually increased during the calculation and depends upon the current time step, the highest current diffusion coefficient and the element thickness.

b) For each element, starting from the exposed face, the bound water, evaporable water, relative humidity, porosity and diffusion coefficient are calculated.

c) The time step counter is increased to initiate the next iteration. The maximum number of time steps is normally set at 96.

3.8 Output Data

Typical output data are illustrated in Table 1 for the indicated input parameters and time step. This type of output has been used to study the effects of cement hardening rate, pulverised fuel ash, mix proportions, curing time and drying path length upon certain aspects of durability. The porosity and bound water contents in the surface region (i.e. up to about 40 mm from the exposed face) are of particular interest since they affect the ingress of aggressive fluids and the environment adjacent to any reinforcing steel.

4 COMPARISON WITH EXPERIMENT

Comparing predictions with experiment is an essential stage in the development of a multivariable model of moisture content and microstructure. If each major variable can be adequately represented for selected experimental conditions confidence is raised to a level where use of the model outside the experimental range is justifiable.

Distance from exposed face (mm)	Relative humidity	Bound water (kg/kg cement)	Porosity (> 4nm wide)	Evaporable water (kg/m^3 of paste)
0	0.650	0.139	0.346	139
10	0.770	0.177	0.254	243
20	0.814	0.185	0.224	284
30	0.833	0.187	0.212	301
40	0.841	0.189	0.206	310
60	0.852	0.190	0.199	319
90	0.859	0.191	0.195	326

Time step - 83
Drying time - 245.9 days   Curing time - 7 days
Cement - 30% pulverised fuel ash + 70% ordinary Portland cement
Water/cement ratio - 0.47
Aggregate/cement ratio - 5.0
Aggregate absorption - 0.5%
Drying path (= $\frac{1}{2}$ section thickness) - 100 mm
Exposure relative humidity - 0.65

TABLE 1 TYPICAL OUTPUT DATA FOR THE INDICATED INPUT PARAMETERS AND TIME STEP

The four, major variables were measured as follows:

a) Relative humidity. A commercial probe was used to monitor the relative humidity in cavities that were cast into a 100 x 125 x 200 mm block of concrete. The block was sealed to allow uniaxial drying from the two 100 x 125 mm faces. Cavities were cast at 20, 45 and 90 mm from the exposed faces. The relative humidity probe was calibrated before and after each set of measurements using five, standard, saturated salt solutions (5).

b) Porosity, bound water and evaporable water. Slices of hydrated cement paste, 3 mm thick, were stored in the cavities in the concrete block so that their exposure simulated that of the local cement paste phase in the concrete. The parent slab of hydrated cement paste, from which the slices were sawn was cast on the same day, with the same cement and with the same water/cement ratio as the companion concrete block. At appropriate ages slices were removed from the cavities for methanol exchange and methanol adsorption measurements. Procedures followed those described previously (4) and allow the total, small (< 4 nm wide) and large (> 4 nm wide) pore volumes to be calculated. The water loss from 105 to 900°C was determined after completion of the porosity measurements as a measure of the chemically bound water content. Certain slices of hydrated cement paste were periodically weighed and then returned to their cavities: these weights, together with an allowance for the increase in bound water content, were used to calculate the evaporable water contents.

Tables 2, 3 and 4 show details of two experiments and some of the comparisons obtained at ages of 20, 50 and 250 days. Relative humidities were reasonably well predicted at later ages but predictions were consistently low at early ages. Bound water content predictions were generally quite good for both experiments, the difference from experimental values being little more than experimental scatter. Porosity (> 4 nm wide) predictions were quite close to experimental values except for the later stages of experiment 1. Evaporable water contents were better predicted in experiment 2 than in experiment 1.

Parameter	Experiment 1	Experiment 2
Ordinary Portland cement (%)	100	70
Pulverised fuel ash (%)	0	30
Water/cement + pfa ratio	0.59	0.47
Aggregate/cement ratio	6.0	5.0
Aggregate absorption (%)	0.5	0.5
Drying path (mm)	100	100
Exposure relative humidity	0.65	0.65
Curing time (days)	7	7

TABLE 2 DETAILS OF EXPERIMENTS 1 AND 2

Age (days)	mm from exposed face	Relative Humidity		Bound water (kg/kg cement)		Porosity (> 4nm diameter)		Evaporable water (kg/m^3)	
		Exp	Pred	Exp	Pred	Exp	Pred	Exp	Pred
20	20	.98	.92	.192	.187	.310	.317	419	399
	45	.99	.95	.195	.190	.310	.303	431	443
	90	1.00	.98	.198	.192	.310	.290	-	488
50	20	.95	.88	.206	.198	.270	.306	389	346
	45	.97	.91	.208	.203	.270	.292	408	379
	90	.99	.92	.210	.205	.270	.285	-	396
250	20	.81	.82	.225	.212	.256	.293	256	293
	45	.84	.85	.226	.216	.250	.280	250	280
	90	.84	.87	.227	.218	.230	.275	230	275

TABLE 3 EXPERIMENTAL (EXP) AND PREDICTED (PRED) DATA:- EXPERIMENT 1

Age (days)	mm from exposed face	Relative Humidity		Bound water (kg/kg cement)		Porosity (> 4nm diameter)		Evaporable water (kg/m^3)	
		Exp	Pred	Exp	Pred	Exp	Pred	Exp	Pred
20	20	.98	.88	.162	.160	.265	.267	367	381
	45	1.00	.96	.162	.163	.265	.250	387	413
	90	1.00	.99	.162	.165	.265	.238	402	460
50	20	.95	.88	.176	.171	.237	.249	326	326
	45	.98	.91	.178	.175	.235	.230	358	357
	90	1.00	.92	.179	.177	.232	.221	374	372
250	20	.79	.81	.191	.185	-	.224	274	284
	45	.84	.84	.193	.189	-	.204	324	313
	90	.86	.86	.194	.191	-	.195	345	326

TABLE 4 EXPERIMENTAL (EXP) AND PREDICTED (PRED) DATA:- EXPERIMENT 2

Data from other experimental conditions are being collected so full judgement of the success of the modelling should be postponed until they are available. However the first two experiments suggest that the model is generally realistic. Also if any consistent discrepancies are observed it may be possible to amend some of the equations and obtain improved predictions.

5 APPLICATIONS

It is often true that further experimentation and refinement of a model will lead to desirable improvements but it is equally true that engineering judgements must be made at a given time even if these improvements are not available. The model described here has been used for several purposes, one of which will be described by way of an example.

5.1 Cements Extended with pfa

Extended cements, in which pulverised fuel ash, pfa is blended with Portland cement, are sometimes specified as an alternative to pure Portland cements and the engineering consequences of such a decision must be assessed. Even with a highly reactive pulverised fuel ash the extended cement is likely to react more slowly than the Portland cement alone. However it will often be possible to reduce the water/cement ratio of a pfa blended cement concrete while maintaining the same workability. Also the pfa will react with the calcium hydroxide produced by the Portland cement reaction, if there is sufficient moisture, and produce a voluminous, calcium silicate hydrate gel that greatly reduces the volume of large pores. The model was used to examine the likely porosity of the surface region of concrete when 30% of the Portland cement was partially replaced by a high quality pfa. Comparisons with the Portland cement concrete were made at the same water/cement ratio and at a reduced water/cement ratio. The latter corresponded to a mix that would have a workability and a 28 day cube strength similar to those of the Portland cement mix. The predicted porosities given in Table 5 correspond to ages when the relative humidity at 20 mm from the exposed face had dropped to 0.8. Once this condition is reached further hydration of the surface region will be minimal and relative porosity values are unlikely to change. It can be observed that direct substitution of pfa for Portland cement increases the porosity of the surface (20 mm deep) region by 10%, but if allowance is made for the water reducing potential of the pfa the porosity of the surface region is decreased by 9%. Thus the predicted effect of blending 30% pfa with Portland cement upon the predicted physical structure of the surface region is not large. However the porosity of the interior region can be significantly reduced due to the slow rate of drying and continuation of the pozzolanic reaction. This self-curing potential of cement could also be realised in the surface regions if an efficient curing

Cement OPC %		100	70	70
pfa %		0	30	30
Water/cement		0.59	0.59	0.49
Days to 0.8 relative humidity at 20 mm from exposed face		218	206	164
Porosity (> 4 nm wide) at indicated distance from exposed face	0 mm	0.434	0.473	0.410
	10 mm	0.334	0.374	0.307
	20 mm	0.291	0.320	0.245
	30 mm	0.268	0.290	0.212
	90 mm	0.235	0.250	0.171
Average porosity 0, 10, 20 mm (Relative values)		0.353 (1.00)	0.389 (1.10)	0.321 (0.91)

Drying path = 100 mm
Exposure relative humidity = 0.65
Curing time = 3 days

TABLE 5 PREDICTED EFFECTS OF BLENDING PULVERISED FUEL ASH WITH PORTLAND CEMENT

membrane was used. It should be emphasised that while porosity may provide a useful indicator of physical structure, chemical factors may also be important. For example, if we consider steel corrosion consequent upon carbonation, the benefit of a slightly lower porosity in the surface region is countered by lower calcium hydroxide levels when pfa is used. In fact some unpublished measurements, corresponding to the Portland cement mix and Portland cement/pfa mix at 0.49 water/cement ratio in Table 5, show that the two effects counterbalance each other and the carbonation fronts advance at similar rates.

5.2 Carbonation

The model does not provide any estimate of the depth of carbonation. However it is known that carbon dioxide diffusing into concrete will react with the cement hydrates and reduce porosity (6,7). Figure 3 illustrates some recent measurements of porosity and diffusion on a carbonating paste in which 30% of Portland cement has been replaced with pfa (8). The carbonation front had penetrated 6 mm and 2 mm into the left and right surface regions respectively after 163 days exposure. This and other experimental data suggest that the model should be extended to include carbonation and its effects upon surface porosity.

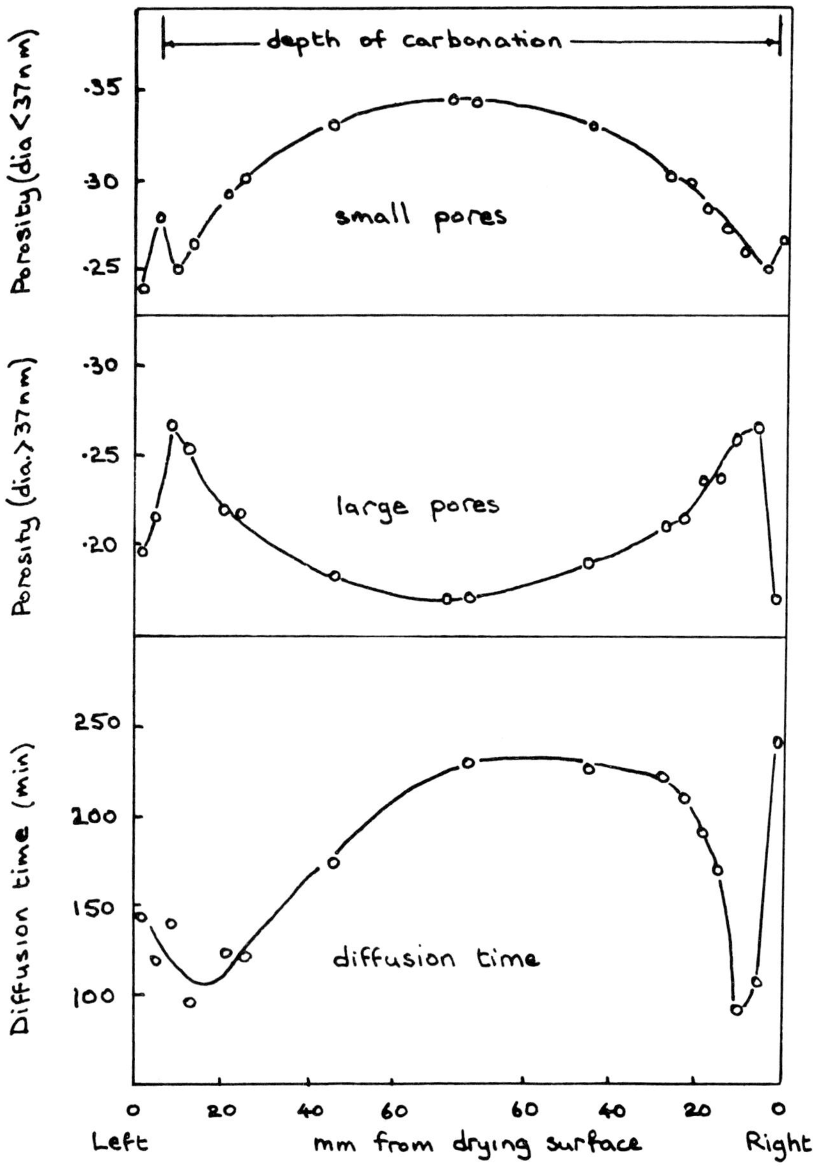

FIGURE 3 CARBONATION, POROSITY AND DIFFUSION DATA VS. DISTANCE FROM DRYING SURFACE (8)

Apart from its effects upon porosity and diffusion, carbonation is also directly relevant to the serviceability of a concrete structure. If the carbonation front reaches the reinforcement then passivation of the steel surface will be destroyed and the reinforcing bars will be susceptible to corrosion. Thus extension of the model to yield the penetration rate of the carbonation front would allow estimation of a service life based on the initiation of carbonation-induced corrosion. The effective depth of cover used in such a calculation ought to allow for the variability of steel fixing and the locally increased depth of carbonation associated with the coarse aggregate/matrix interface.

Details of how the model will be extended to include carbonation have not yet been finalised but the broad principals are as follows:

a) The controlling reaction is $Ca(OH)_2 + CO_2 = CaCO_3 + H_2O$

b) The reaction rate is assumed to be controlled by CO_2 and water availability.

c) CO_2 availability is controlled by surface concentration, porosity of the carbonated layer, moisture in the carbonated layer and reaction of CO_2 with the calcium silicate hydrate gel.

d) $Ca(OH)_2$ availability is controlled by the amounts of cement and pfa reacted at the carbonation front.

e) The rate of penetration of the carbonation front is jointly controlled by the amount of $Ca(OH)_2$ to be carbonated and the supply of CO_2.

f) The formation of $CaCO_3$ reduces porosity.

g) The water released increases the evaporable water content.

5.3 Design for Durability on the basis of porosity

The model could be used in its present state to study the effects of exposure relative humidity, curing time, section thickness, mix proportions and cover to the reinforcing steel upon the average paste porosity, p of the concrete between the exposed surface and the steel. This porosity data can then be combined with the depth of cover, cx to calculate a diffusion resistance parameter for the concrete cover. If, to ensure durability, a given severity of exposure is associated with a particular diffusion resistance parameter, a consistent approach to the choice of mixes, curing and cover could be developed for use in engineering design. On the basis of simple diffusion theory the diffusion resistance parameter might be quantified as cx^2/p^4, although other formulations, perhaps including a moisture state component, are possible.

General application of this method is probably best deferred until additional comparisons between predictions and experiment have been made and the effects of carbonation upon surface porosity have been incorporated: favourable comparisons would significantly increase confidence in the validity of the model and its use in engineering design.

6 CONCLUDING REMARKS

The exposed surface layers of concrete can be significantly different in microstructure from the interior region, where moist conditions and prolonged curing occur. It is these surface layers that, in spite of their lower quality, have to protect the interior concrete and reinforcing steel from the ingress of harmful fluids. The model described in this report was developed in an attempt to quantify these variations in microstructure. Although it contains many assumptions, initial results suggest that the model will improve our understanding of how engineering parameters such as mix proportions, curing, depth of cover etc., can affect durability. Additional development of the model to include carbonation and its effect upon microstructure is considered essential. Experiments are now in progress to establish the reliability of the model.

REFERENCES

1. Parrott, L.J. and D.C. Killoh. Prediction of Cement Hydration. To be published in Proc. Brit. Ceramic Soc.
2. Powers, T.C. and T.L. Brownyard. Studies of the Physical Properties of Hardened Portland Cement Paste. Portland Cement Assoc. Res. Dept. Bull., No. 22 Chicago (1948).
3. Parrott, L.J. Novel Methods of Processing Cement Gel to Examine and Control Microstructure and Properties. Phil. Trans. R. Soc. Lond. A 310 (1983) 155-166.
4. Parrott, L.J. Effect of Drying History upon the Exchange at Pore Water with Methanol and upon Subsequent Methanol Sorption Behaviour in Hydrated Alite Paste. Cement and Concrete Res. 10 (1981) 651-658.
5. Young, J.F. Humidity Control in the Laboratory using Salt Solutions - a Review. J. App. Chem. 17 (1967) 241-245.
6. Jungermann, S. The Chemical Process of the Carbonation of Concrete. Betonwerk + Fertigteil-Technik 6 (1982) 358-362.
7. RILEM Int. Symp. Carbonation of Concrete. Cement and Concrete Assoc., Slough (1976).
8. Patel, R.G., L.J. Parrott, J.A. Martin and D.C. Killoh. Gradients of Microstructure and Diffusion Properties in Cement Paste Caused by Drying. To be published.

MATHEMATICAL MODELLING OF ATMOSPHERIC CORROSION AND ENVIRONMENTAL FACTORS

Svein E. Haagenrud

Norwegian Institute for Air Research
P.O.Box 130, N-2001 Lillestrøm, Norway

ABSTRACT

The corrosivity of the atmosphere is described by so-called dose-response (D/R) relationships between environmental variables and corrosion rates. Laboratory and field studies by NILU to establish such relationships for carbon steel and zinc are described. These functions can be derived for local-, meso- and microclimatic regions on short (monthly) and long term (years) basis. Standardization work on international level uses D/R-functions to quantitatively classify the atmospheric corrosivity. Dispersion modelling of air pollution can be used to estimate the environmental variables in the D/R functions, allowing for a surveillance program for corrosion without extensive corrosion rate measurements.

1 INTRODUCTION

Atmospheric corrosion is of great practical importance and accounts for more than 50% of the total cost caused by corrosion. Since sulphur pollutants play an important role in increasing the corrosivity of the atmosphere, the increased pollution in many parts of the world has in the last three decades increased the atmospheric corrosion aggressivity.

Knowledge of the corrosivity of the atmosphere would permit a reduction of the damage by an appropriate choice of material, surface protection and maintenance measures. Another way to decrease the corrosivity would be to reduce emissions. In Europe the ECE member countries thus agreed to reduce emissions (1). The problem of atmospheric corrosion as well as other damages from sulphur pollutants is also of great concern in the US (2).

It is generally recognised that the processes of corrosion and degradation of function are complex phenomena, governed by many factors. The relationship between these factors and corrosion are often expressed by so called dose/response (D/R) relationships. The development of such functions are a key subject of work in fighting corrosion. Field studies are necessary to establish these functions, and this has been the subject of several national and international research programs. There is at present considerable experience in the area of atmospheric corrosion of metals. In some countries standards or codes of practice including a classification system of the corrosivity of the atmosphere have already been adapted.

This paper will discuss some of the important variables in the mechanism of atmospheric corrosion, and some of the work performed to establish necessary D/R relationships. The development, use and limitation of such functions for prediction of service life is discussed, as well as the structure of the system for classification of atmospheres.

2 THE MECHANISM OF ATMOSPHERIC CORROSION

The mechanism of atmospheric corrosion is dependent on the material involved, the atmospheric climatic conditions and pollutants present. Apart from the pollution level there are many variables which will influence the final corrosion rate. This includes relative humidity, temperature, rainfall, windspeed and local topography of the site, in addition to the physical characteristics of the exposed material such as shape and heat capacity.

2.1 Climatic factors

For metals the process is electrochemical in nature, requiring a film of moisture. In practice this results in a discontinuous process and means that the time during which a moisture film is present on the surface is a major factor in the final observed corrosion rate. This is often called the time of wetness (TOW). Assessment of TOW has been used in all modern survey work either by direct measurements (3) or by relating this parameter to the length of time the relative humidity is above 80-90%. The true TOW on the surface depends on the corrosion products and the air pollutants present.

Precipitation can act in two ways. For small amounts and short duration, the rain will wet the surface and promote corrosion. Large quantities and/or long duration will reduce corrosion to some extent by washing pollutants from the surface, but removal of surface layers affording protection to the material will result in increased corrosion. The acidity of the rain also may increase the corrosion rate (4).

Wind speed could also have a dual effect on corrosion rates. High wind speed can have a drying effect on the surface and corrosion rates will decrease. On the other hand, increasing wind speeds could enhance the supply rate of SO_2 and thus increase the corrosion rate (5).

Temperature has been found to be significantly correlated to corrosion rate in several studies. On the other hand, high temperature may also dry the surface, thus decreasing the overall rate (6,7).

2.2 Influence of sulphur compounds

Sulphur emissions consist primarily of sulphur dioxide, whose effect on material damage has been well demonstrated in both field and laboratory investigations. Sulphur containing gases and aerosols accumulate on the exposed surfaces of materials by both dry deposition - the direct absorption of gases and the sedimentation of aerosol particles, and by wet deposition - precipitation (8). Many corrosion studies have been undertaken measuring only SO_2, making the assumption that SO_2 is the main deteriorating agent. This is one of the main criticisms raised against many of the D/R studies performed. However, studies with more extensive measurements have shown only weak correlations between sulphate particulates and corrosion rates (9).

Several studies have shown that H_2S in the concentrations usually present in the atmosphere causes only limited corrosion of the most important materials in outdoor atmospheres. There is, however, evidence that H_2S is an important pollutant in indoor atmospheres, contributing especially to the corrosion of electric equipment and contact materials (10).

2.3 Influence of other pollutants

Other non-sulphur pollutants which are present in the atmosphere include nitrogen oxides, particulates, chloride, ozone and impurities in rain. However, little quantitative work has been carried out to determine the effect of these pollutants either alone or in combination.

With regard to nitrogen oxides or nitrate particles most of the published data indicates that NO_x is of limited importance for acceleration of the outdoor corrosion. Even so, more comprehensive investigations which take into account possible synergistic effects with other pollutants will be useful.

An area where the influence of NO_x is of special interest is the corrosion of electric contact materials. Laboratory studies have shown that nitrogen oxides in combination with other air pollutants, such as SO_2, have a corrosive effect on contact materials (11).

There are two main types of atmospheric chloride contaminant, sea-salt and HCl gas. The major sources of HCl include chemical plant, municipal waste incinerators and combustion of chloride-containing coal. Small particles of sea-salt may be transported over very large distances but, in general, the major effect on corrosion is within some hundred meters of the sea-shore. It promotes corrosion by the formation of an electrolyte on the surface at low relative humidities (50-60% RH), thus effectivly increasing the TOW. Similarily for HCl the TOW will also be increased. The low pH level will also increase the corrosion rate, as will the general corrosive action of the chloride ion itself.

Rain acidity can cause dissolution of the protective layer and promote corrosion. The effects of ozone, hydrocarbons and carbon monoxide have been investigated in only a few studies (6). Direct corrosive effects of ozone on paints, elastomers and plastics are well known (12).

3 DEVELOPMENT OF DOSE/RESPONSE RELATIONSHIPS

The effects of the environment on the corrosion rate of various materials are usually studied by exposing panels at test sites in various types of atmospheres (Figure 1). When both the corrosion rate and the environmental variables are properly measured D/R relationships may be developed by mathematical analysis. Numerous such investigations have been made and reported in the literature (1). Climatic chamber tests are often used to study the effect of one or more parameters under controlled conditions.

In the following sections some of the work done by NILU on a national and international level will be described. Most of this work has been previously published.

Figure 1: A typical outdoor exposure rack.

3.1 Effects of Long-Range Transport of Air Pollutants (LRTAP)

Sulphur pollutants transported over long distances are deposited over southern Norway to such an extent that increased atmospheric corrosivity may be expected (13). Field and laboratory investigations were therefore undertaken to study this possibility.

3.1.1 Climatic chamber study of pH effects. In a climatic chamber, the effects of various pH values in precipitation on the corrosion rate of carbon steel and zinc have been studied (4). The tests are performed as four week tests, the corrosion rates being determined both as weight-loss of panels and with electrolytic cells (4). The test program with a total TOW of about 55 hours for the test period, and the results are shown in Figure 2. Zinc specimens have been pre-exposed for three days under defined humidity conditions to give a stable surface film (14).

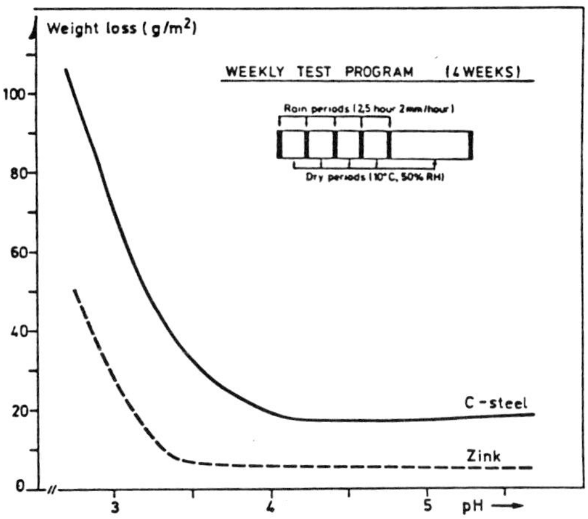

Figure 2: Corrosion rate of carbon steel and zinc as function of pH in precipitation.

The results show that for carbon steel there was no effect of decreasing pH in the range of pH 5.7-4.2 (threshold level), but decreasing of pH to below 4 sharply increased the corrosion rate, At pH 2.7, the corrosion rate was about 7 times higher than at pH 5.7- 4.2. The D/R relationship seems to follow an exponential function from the threshold level.

The corrosion rate of zinc was less dependent on pH than that of carbon steel, contrary to what is seen in water. It seems to to be more or less constant in the pH range 5.7-3.7 (threshold level), whereas decrease of the pH to below 3.7 seems to give a sharp increase in the corrosion rate. The D/R relationship seems to be approximately the same function as for steel.

3.1.2 Field studies. This program involved monthly measurements of environmental variables and corrosion rates of carbon steel and zinc at the main LRTAP test site Birkenes for a 25 month period (Fig. 3).

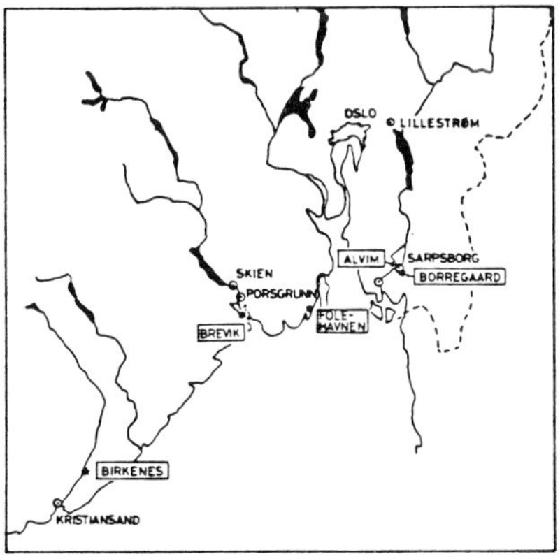

Figure 3: Corrosion test site Birkenes-geographical location.

Parallel to the monthly program corrosion was measured after 1, 3, 6, 12, 24, 36 and 60 months of exposure.

Detailed analysis of the exposure of weight-loss coupons is described elsewhere (15).

The conclusions from this study were that the monthly corrosion rate of carbon steel as measured by weight-loss at Birkenes, is influenced by long-range transport of air pollutants occurring as episodes of acid precipitation with pH less than 4 and high SO_2 concentrations. The monthly corrosion rate for steel is thus best described by the following relation.

$$K_{Fe} = 2.3\ (M\ S\ 95) + 1.0\ SO_2 + 0.01\ Cl + 0.1 \qquad (1)$$

with a correlation coefficient $R = 0.9$

where K_{Fe} = corrosion rate (g/m^2 month)

(M S 95) = (amount of precipitation with pH <4) X (hour > 95% RH)

SO_2 = sulfur dioxide concentration in air($\mu g/m^3$)

Cl^2 = deposition of chloride in precipitation (mg/m^2)

Figure 4 show measured and calculated corrosion rates of carbon steel after this equation. The impact of the episodes in Nov. 74 is evident.

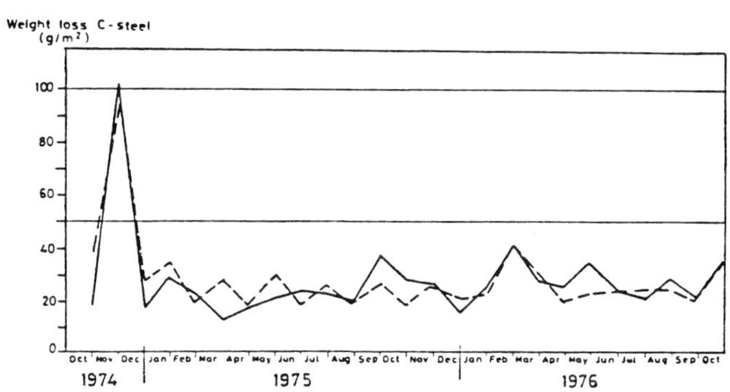

Figure 4: Observed (——) and calculated (--) monthly corrosion rates of C-steel after equation 1.

These episodes, however, seem to occur so seldom that the corrosion rates measured for longer periods of time do not seem to be influenced, i.e. for carbon steel the corrosion agressivity at Birkenes for that given period must be classified as typical for an average rural atmosphere.

For zinc the monthly corrosion rates vary considerably and are often high. This is in accordance with previous findings (14) and is mostly caused by variations in the TOW during the first one to seven days of exposure. Regression analysis showed that the monthly corrosion rate was far better correlated with "hours of precipitation" during the first days of exposure that with the total "hours of precipitation" per month. Monthly and yearly corrosion rates gave no significant evidence of any effect of LRTAP on the corrosion rate of zinc measured on standard weight-loss coupons for this period of time. However, for zinc the corrosion aggressivity at Birkenes must be classified as a relatively aggressive rural atmosphere, which is probably due to the large amount of precipitation (See 5.1).

By knowledge of the environmental data, equation (1) could be used to predict the monthly corrosion rates and the influence of the long range transported air pollutants. Such data can be extracted from the LRTAP model developed in the OECD-study for Europe (16).

3.2 Estimating Long Term Corrosion Rates from BIALBO-model

This program include the Birkenes study described above, together with additional test sites at Alvim and Borregaard, which can be classified as urban (20-40 μg SO_2/m^3 month) and industrial atmosphere (about 80-100 μg SO_2/m^3 month), respectively (Figure 3).

Figure 5 shows that the corrosion rates for steel for every month and for a long term exposure are by far highest at the industrial site, except for Nov. 74 at Birkenes, as explained in 3.1. The correlation between the monthly corrosion rate for carbon steel at all three sites and each single environmental variable indicates that the corrosion rate is well correlated to sulphur dioxide (R = 0,84) as shown in Figure 6 and strong acid (R = 0.51). These correlations are significant at the 99.5% confidence level.

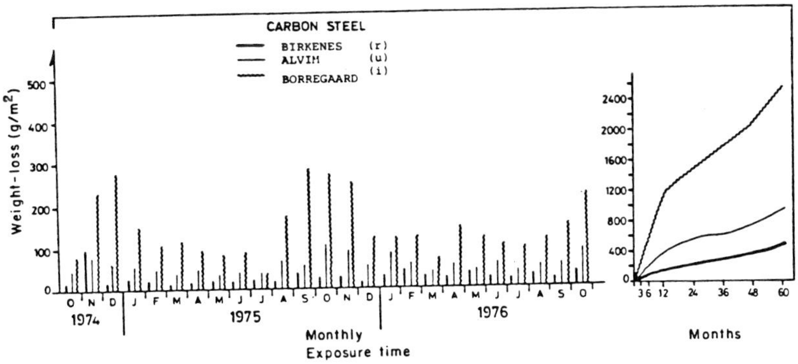

Figure 5: Corrosion of carbon steel by monthly and 3, 6, 12, 24, 36, 48 and 60 months exposure at the three test sites: <u>BI</u>rkenes, <u>AL</u>vim and <u>BO</u>rregaard (BIALBO).

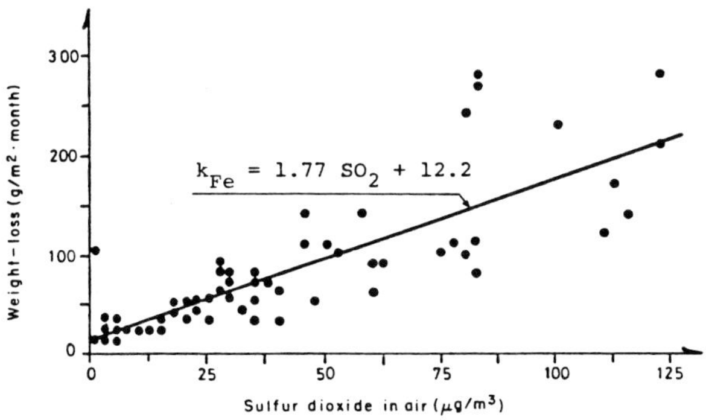

Figure 6: Monthly corrosion of carbon steel and sulfur dioxide in air at the three test sites Birkenes, Alvim and Borregaard.

The correlation coefficients changes considerably for each site, indicating that the corrosion determining parameters may be different in different types of atmosphere. In multiple linear regression analysis the best equation taking the sum of 2, 3 and 4 parameters have been calculated. The best equation obtained is

$$K_{Fe} = 1.54\ SO_2 + 2.34\ DPREC + 0.05\ H^+ - 15.2 \qquad (2)$$

with $R = 0.87$

where

K_{Fe} = monthly corrosion rate of carbon steel (g/m^2)

SO_2 = concentration of SO_2 in air ($\mu g/m^3$)

DPREC = days with precipitation

H^+ = concentration of strong acid in precipitation (μ-equiv. H^+/l)

Using this equation, called the BIALBO model, the monthly corrosion rates of carbon steel can be calculated at other locations where the environmental parameters are known or can be estimated. These calculated corrosion rates can then be compared with the monthly corrosion rates from these three sites where long term corrosion rates are available (Fig. 5). The yearly corrosion rate at the location in question can then be calculated or evaluated with a fair degree of accuracy.

Recently the Sarpsborg/Fredrikstad urban area, where Alvim and Borregaard are situated (Figure 3), has been the subject of a thorough case study of modelling of emissions, dispersions and effects of air pollutants. This study paid for by the State Pollution Control Authority has given the opportunity to test out the developed D/R model (Eq. 2) (see Chapter 4).

3.3 Macroclimatic Corrosion Mapping

In order to map the corrosivity of the atmosphere in Scandinavia and to find relations between corrosion and environmental variables, a common Nordic research programme was started in 1975 at 32 test sites in different types of atmospheres (Fig. 7).

As a more detailed description of the test conditions and the results has been given earlier (17,18), only the main features will be presented here. The results will be further discussed in Chapter 5.

The materials tested are carbon steel and zinc and the exposure periods are 1, 2, 4 and for carbon steel 8 years.

To make the regression analysis meaningful the sites were grouped into rural, sulphur dioxide, and chloride polluted, respectively. This was done according to the ISO classification criterion proposed (see Chapter 5).

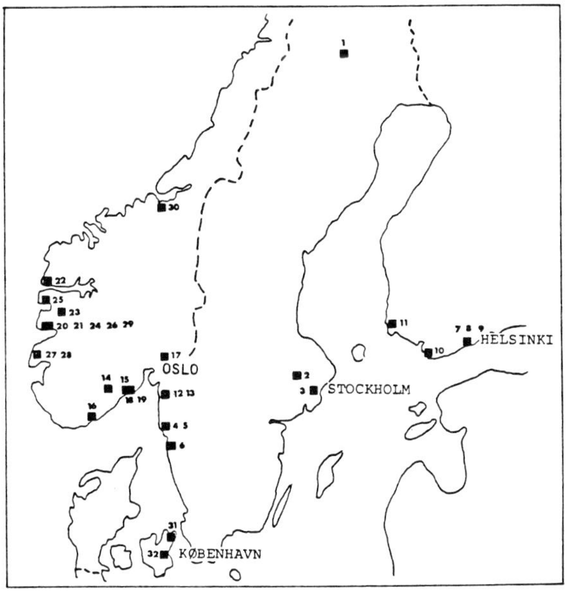

Figure 7: 32 Corrosion test sites in Scandinavia.

The regression lines for corrosion rates and SO_2 are shown in Table 1.

Table 1: Regression lines for influence of SO_2 on corrosin of carbon steel and zinc at 1- and 4 years exposure in Scandinavia.

	Regression lines 1 year (18)	4 years	Corr. coeff.	Unexpl. variance (%)
(3)	$K_{Fe}=5.28\ SO_2+176.6$		0.76	42
(4)		$K_{Fe}=18.5\ SO_2+292.5$	0.88	22
(5)	$K_{Zn}=0.22\ SO_2+6.0$		0.76	42
(6)		$K_{Zn}=1.1\ SO_2+14.6$	0.88	22

The 4-year regression lines are shown in Figure 8a and 8b. As seen the corrosion rates are highly correlated with the SO_2-concentration, the 4-year exposure having the highest correlation. From a comparison of the constants in the equations it may be concluded that SO_2 affects the corrosion rate in the same way for both materials while the background corrosion based on 4 years' data is approximately half of that based on 1 year's exposure. Figure 8a and b also show that the SO_2 content from the industrial Borregaard site has a strong influence on the correlation.

It has thus to be concluded that the sites are not optimally localized for determination of more accurate dose response relationships. This would demand sites with a greater dispersion of the SO_2 concentrations.

The influence of other variables, as well as the possible error in data, are expressed by the unexplained variable. Due to missing data from certain sites and in several cases due to small variations within the region of Scandinavia, as well as to a certain lack of quality it is difficult to evaluate the influence of other variables.

Comparison of the regressing equations in Table 1, with regressions obtained in a similar study for the Swedish sites and 5 Czechoslovak test sites (19) show a marked similarity. For zinc comparison may also be done with a study by Hudson and Stanners in UK: (21).

$K_{zinc} = 0.22\ SO_2 + 6.0$ (5)
$K_{zinc} = 0.21\ SO_2 + 4.5$ Knotkova et al. (20).
$K_{zinc} = 0.2\ SO_2 + 3.1$ Hudson & Stanners (21).

The agreement between the three equations may be considered very good.

The D/R relationships between corrosion rates and chloride for 4-year exposure were:

(7) $K_{Fe} = 364.1 \log (Cl) + 279$ with $R = 0.89$, $V = 20\%$
(8) $K_{Zn} = 15.1 \log (Cl) + 15.9$ with $R = 0.91$, $V = 17\%$

which are statistically significant at 99% level. Two remarks should be made to these relations: Firstly, the correlation depends very much on the results from one sea shore site, and secondly, the chloride concentration (Cl-mg/l) is measured by a rain collector. In the future a comparison with data obtained by another sampling method, e.g. the Ambler wet candle, would be very useful.

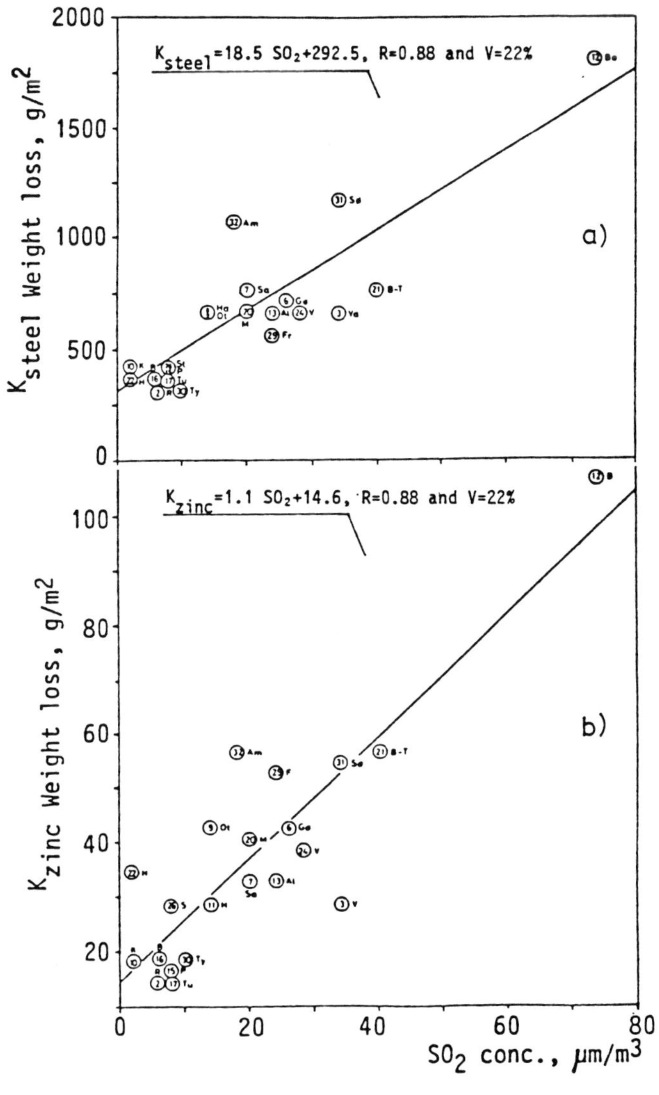

Figure 8: The corrosion of carbon steel and zinc at 4 years exposure at test sites in Scandinavia as function of the SO_2 content (16).

The Scandinavian mapping study permit the following main conclusions:

- Exposure of steel and zinc specimens at 32 sites in Scandinavia shows that the pollution by SO_2 in urban and industrial areas and by chlorides in marine locations is of decisive importance for the corrosion rate.
- Comparison with similar studies in Central Europe and UK show that the D/R relationships may be applicable for an even wider geographical area.
- The results may serve as a basis for a quantitative description of corrosivity classes using SO_2 concentration and chloride deposition, which will be of great practical importance for standardization purposes (Chapter 5).
- Also the corrosion rates from 1 year's exposure of standard specimens seem to give a useful indication of the corrosivity of the atmosphere.

4 DISPERSION AND CORROSION MODELLING OF AN URBAN AREA

4.1 Dispersion Modelling

The Sarpsborg/Fredrikstad area is a densily populated and industrialized urban area. An emission inventory of air pollutants have been carried out, and Figure 9 shows the emission of SO_2 for the area in grids of 1 km x 1 km. The total SO_2 emissions in 1981 were about 8900 tons. Borregaard Industries with about 60% of the total emissions were the greatest emitter, while the total industrial emissions represented more than 90% of the total.

Shown in Figure 10 are the calculated mean SO_2 concentration profile for the Sarpsborg/Fredrikstad area for the period November 1981- March 1982. The model used is NILU's standard Gaussian dispersion model modified to account for turbulence generated by buildings. Results from tracer gas dispersion studies have been used to describe dispersion from low level emissions (23).

There is fair consistency between calculated and measured SO_2 concentrations at fifteen measuring points in the area. Measured concentrations are also indicated on Figure 10. The model also allows for calculations of the relative contributions from various emission sources. The results of this calculations are shown in the upper part of Figure 10. It indicate that the greatest contributors by far to the ambient SO_2 concentrations are the emissions from the low industrial

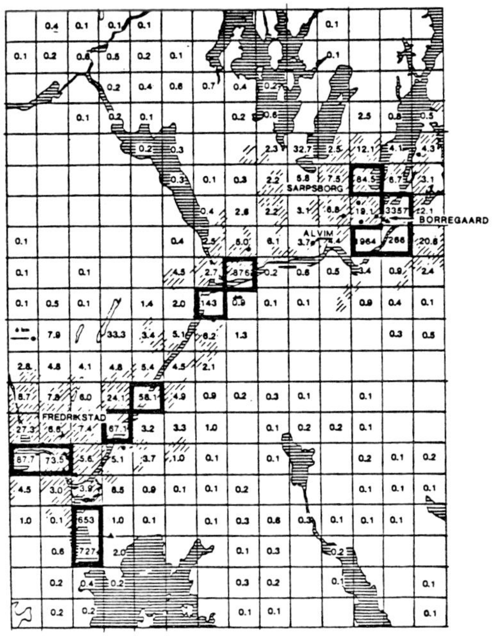

Figure 9: Total emission of SO_2 in Sarpsborg/Fredrikstad area 1981 (ton/year).

stacks. These emissions are in the turbulent air zone around the buildings. However, these emissions account for only about 10% of the total emissions in the region, while the greatest industrial emissions accounting for about 80% of the total emissions only contribute with maximum 10 $\mu g/m^3$ in mean SO_2 concentrations on the ground. This is due to the high emission height.

SO_2 emissions from home heating contribute only about 8% of the total emissions in the area. In Fredrikstad, however, these sources are the main contributors to the SO_2 concentration in ambient air.

The main conclusions from the air pollution modelling are that the dispersion model describes the real occurring air pollution situation with a fair degree of accuracy. The information obtained can be used by the authorities to choose the most appropriate actions for effective improvement of the atmospheric environment in the area.

Corrosion rates have been measured simultaneously at the fifteen measuring points in the area, and D/R functions are being developed. However, being at present in the reporting process only some preliminary results from comparison with the BIALBO monthly model can be given.

SO_2 contributions from:

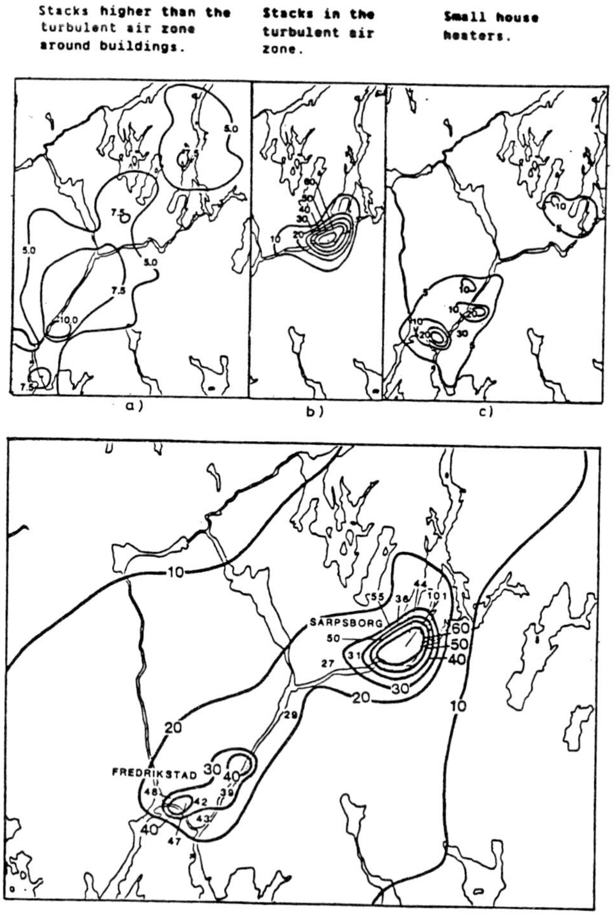

Figure 10: Calculated mean SO_2-concentration in period Nov 81 - Mar 82. Measured concentrations are shown at the measuring sites.

4.2 Comparison with the BIALBO monthly model

The measured monthly corrosion rates of carbon steel in the period Nov 82-Oct 83 have been compared with calculated corrosion rates (Fig. 11) from the BIALBO model (see 3.2).

$$K_{Fe} = 1.54\ SO_2 + 2.34\ DPREC + 0.05H^+ - 15.2 \qquad (2)$$

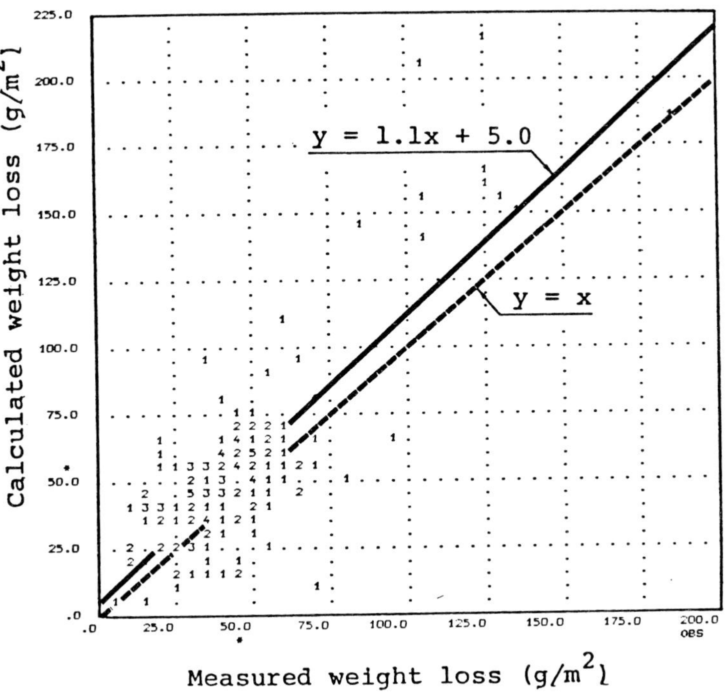

Figure 11: Comparison between 165 monthly corrosion rates - measured and calculated by the BIALBO model - in the Sarpsborg/Fredrikstad area Nov. 82-Sep. 83.

The hypothesis of equal meteorological conditions for the whole area for the same time period has been confirmed (23), and therefore the term DPREC is a constant for all 15 sites for each month. Three levels of (H^+) have been used for the sites, i.e. the rural, urban and industrial. Erroneus corrosion data from Oct 83 have been excluded. It can be seen that the agreement is quite good, the relationship being

$$K_{calc} = 1.1\ K_{meas} + 5.0 \qquad (9)$$

with R = 0.81 and V = 35%.
where K_{calc} and K_{meas} are, respectively, the calculated and measured corrosion rates.

By detailed analysis of the data the corrosion model will be improved. Using that model and the SO_2 dispersion model (Fig. 10) a surveillance program for corrosion in this area as well as in areas with related environments can be established.

5 PRACTICAL APPLICABILITY OF RESULTS

5.1 Standardized classification of atmospheres

In some of the Scandinavian countries codes of practice for the choice of materials and surface protection have been issued, including a classification system of the corrosivity of the atmosphere. The corrosivity classes are, however, defined in a rather qualitative manner.

As the D/R relationships describe the corrosivity of the atmosphere these functions could be used for a quantitative classification. Efforts are at present been made to work out an internationally accepted system for classification of the atmospheric corrosivity within the scope of ISO/TC 156/WG4. This system is based on the classification of the important environmental variables and/or the corrosion rates of various metals (24).

Using the proposed classes of pollution levels of chloride and SO_2, the test sites in Scandinavia have been classified as shown in Table 2 (18). These sites cover the range of TOW 2500-4000 hour per year, using the criterion proposed by ISO/TC 156/TC 156/WG4, i.e. the time with RH > 80% and T > 0^0C (24). The test sites may be divided into three categories according to the ISO proposal: rural sites with <10 µg/m^3 SO_2 and two urban categories, U 1 with 10-75 µg/m^3, and U 2 with >75 µg/m^3.

A comparison of the results from the rural sites with SO_2 levels <10 µg/m^3 shows that the corrosion rate of steel based on 4 years' exposure at 9 sites is within a very narrow range of 11 to 13 µm/year. This shows that the variations in the meteorological parameters affecting the corrosion rate of

Table 2: Classification of corrosivity at test sites in Scandinavia based on the pollution level and on corrosion rate of steel and zinc after 1 and 4 years exposure (19).

Type of atmosphere, designation		Pollution level Cl^- mg/m^2d	SO_2 $\mu g/m^3$	Corrosion rate, μm/year Steel 1 year	Steel 4 years	Zinc 1 year	Zinc 4 years	Site
Rural	R	<3	<10			0.6-0.8	0.5-0.7	2 Ryda, 10 Koski,
				17-24	11-13			16 Birkenes, 17 T
						(1.7-2.0)	(0.9-1.2)	(22 Hyllestad, 23
Urban	U 1	43	10-75	30-39	22-25	0.0-1.9	1.0-1.5	3 Stockholm, 7 Sa
								11 Harjavalta
		7-47	10-75	35-63	19-38	1.3-2.4	1.4-2.0	6 Gøteborg, 20 Mi
								Tidende, 24 Verit
								31 Søborg, 32 Ama
	U 2		75-200	81	57	3.5-3.8	3.8	12 Borregaard
Marine	M 1	3-20	<10	25-35	14-20	0.7-1.4	0.7-1.1	5 BM1, 27 Stord I
								25 Mongstad
	M 2	20-100	<10	36-68	20-35	1.3-2.0	1.1-1.6	4 BM3, 8 Harmaja,
	M 3	>100	<10	86	40	2.1	1.9	19 Folehavnen 10
Subarctic	A	<3	<10	6	4	0.8	0.4	1 Gällivare

* Sites with large amount of precipitation.

steel are small in the southern part of Scandinavia. In the northern part of Scandinavia, however, the corrosion rate of steel in unpolluted atmospheres is only 4 µm/year. This may be ascribed to the long periods with temperatures below 0^0C. Due to this, the TOW is substantially shorter than in southern Scandinavia.

Exposure of zinc gives in principle similar results showing corrosion rates in rural regions of 0.5-0.7 µm/year. The exceptions are the three rural sites on the Norwegian West Coast, which show rather high corrosion (1.7 to 2.0 µm/year) during the first year of exposure. This may be explained by the large amount of precipitation, which affects the corrosion rate of zinc at the beginning of exposure, as shown by Ellis (14). This so called memory effect may exert its influence during the first years of exposure. After 4 years the corrosion rate has diminished and, in the long run, it seems sto reach a value similar to that at other rural sites.

At the SO_2-polluted sites the corrosion rate of both steel and zinc in general have been found to be 2-5 times higher than at the rural sites. In Scandinavia, with its long coast several areas with combined SO_2 and chloride, pollution may be found. In the U 1 category, there are 8 sites with Cl^- pollution - the two Danish sites probably belong to this category even if their chloride deposition has not yet been established. The corrosion rates at these sites, both for zinc and steel, are somewhat higher than in the U 1 category without chloride pollution (Fig. 9). The SO_2 content of the atmosphere seems, however, to be of decisive importance for corrosivity. Borregaard is the only site belonging to the highest U2 class.

The marine sites have been divided into three categories M 1-M 3. The pollution levels are based on values of combined wet and dry deposition in open precipitation gauges in the order of 3-20, 20-100 and >100 $mgCl/m^2$ d. The use of deposition of chloride instead of distance from shore is preferable, because the deposition strongly depends on local topography, prevailing wind direction, etc.

The results thus show that it is possible to define the corrosivity at sites in southern Scandinavia if the pollution by SO_2 and chlorides is known. The qualitative descriptive classification of the atmosphere into rural, urban and marine, mainly used at present, may thus in the future be replaced by a quantitative system which is based on pollution levels which characterize the individual types of atmosphere. This is a substantial advance and the results will thus be useful both in national and international standardization work.

The results show that also the other possibility of classifying corrosivity by means of the corrosion rate from 1 year's exposure of standard specimens of steel and zinc may be practicable.

5.2 Prediction of Long Term Corrosion Rates

Although the atmospheric corrosivity may be classified from short term data, long term corrosion data are often necessary to predict service life in building industry.

As can be seen from Figure 6 and Figure 12b, the integral corrosion (K) as a function of time (t) during the first 4 years of exposure can fairly well be described by the equation

$$K = k\,t^n, \qquad (10)$$

where k and n are constants. From the values of the constant n, the type of evolution of the corrosion rate with time can be roughly estimated. If n>1, the corrosion rate is decreasing with time and, if n ≈ 1, it is practically constant with time.

The corrosion rates for 1, 2 and 4 years exposure of steel at selected sites, according to the classification in Table 2, show a sharp decrease after the first year (see Fig. 12a), i.e. n > 1. On prolonged exposure the decrease is thereafter only slight with the exception of the SO_2 polluted sites Borregaard and Harmaja. The corrosion rate in marine atmosphere is high during the initial period. Thereafter it slows down, and thus there may be changes in the relative corrosivity of site depending on the length of the exposure period used for a comparison. This may be demonstrated by the Borregaard and Folehavnen 10 m sites. The sharp decrease the corrosion rate during the initial period may also be seen from regression lines (Table 1, section 3.3).

For zinc the decrease in corrosion rate after the initial period is much less pronounced (Fig. 12b). In atmospheres polluted with both SO_2 and chloride, the decrease is only slight. Only the two rural sites Ryda and Gällivare show a sharp decrease of the corrosion rate. This is so also for the already mentioned rural sites at the Norwegian west coast.

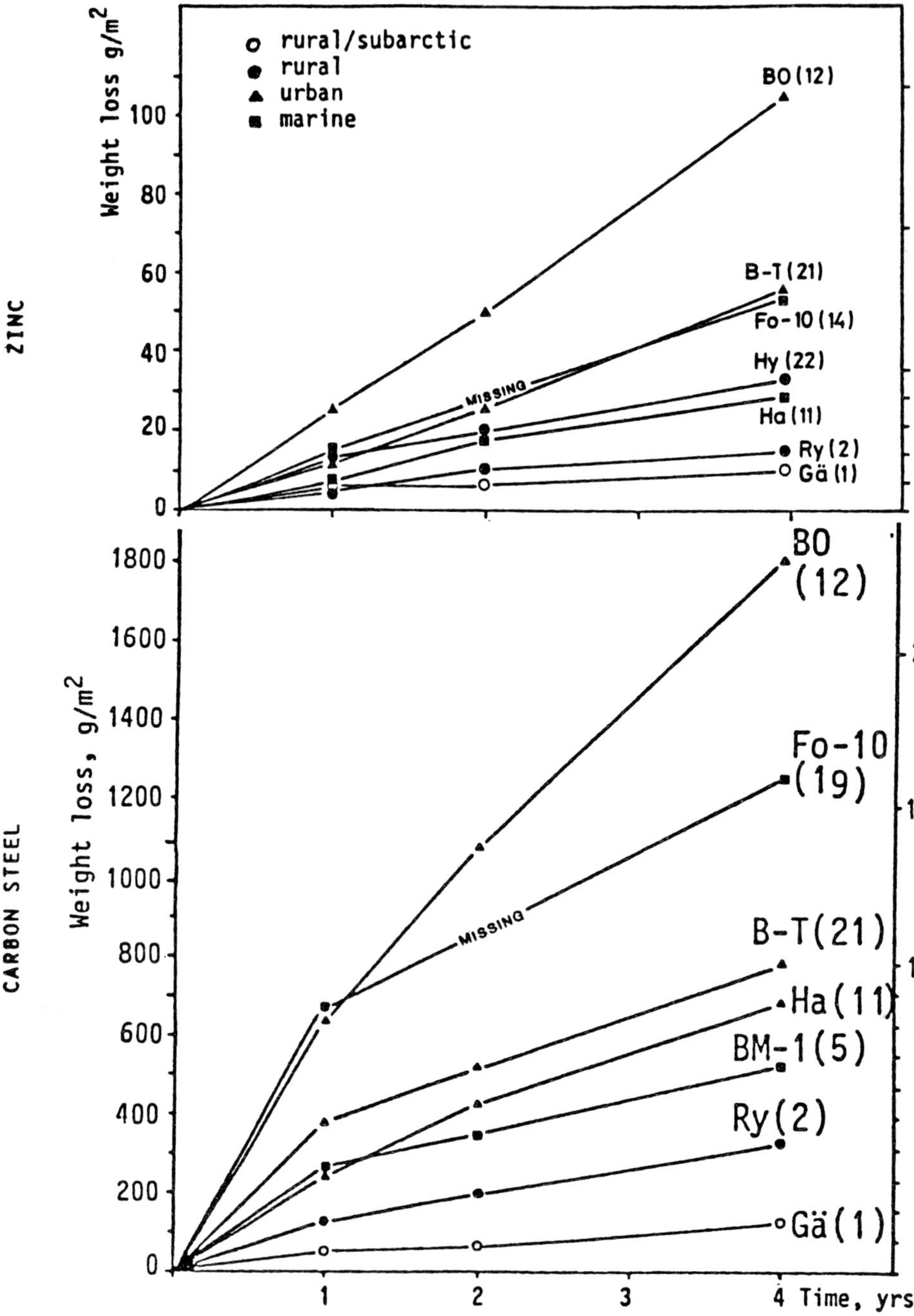

Figure 12: Corrosion of carbon steel and zinc at some test sites in Scandinavia after 1,2 and 4 years exposure.

The expression of the variation of the corrosion with time after only four years are, at present, mainly considered as means for describing of the variations within this exposure period. They could be used as an approximative prognosis for a limited period. The question of prognostic equations for lengthy exposures will be considered after the evaluation of the 8 year results.

5.3 Application to Practical Service Life

When using various D/R relationships in predicting real service life some important points should be remembered. Firstly, it must be realised that these D/R relationships are determined for standardized field test conditions, while the values of environmental variables in question may be strongly influenced by the microclimate. For example, the TOW on different parts of the structure will vary with the degree of sheltering, the orientation in relation to the prevailing winds, solar radiation (drying), etc. The same goes for the SO_2 deposition which is dependent on wind direction, speed, height above the ground, etc.

The influence of the microclimate does not mean that the corrosion mechanism changes, only quantitative values of the relevant variables changes in relation to the values obtained on the greater scale, i.e. meso or macro scale. This implies that for a precise assessment of the environment either measurements or rather sophisticated evaluations are neccessary. Use of dispersion models as described may be of great value in this connection.

Another point to remember is that the material in question is always a part of a component, which in turn is only a part of a complete construction. This may in principle change the corrosion rate and therefore the lifetime of the material in question completely.

6 REFERENCES

(1) ECE document ENV/IEB/R16: Effects of sulphur compounds on materials, including historic and cultural monuments, Sept. 1982.

(2) NAPAP: Materials effects task group (G) and assessments and policy analysis task group (I), peer review meeting, February 1984 Burlington, Vermont. Draft report April 1984.

(3) P.J. Sereda, 1974, Corrosion in Natural Environment, ASTM, STP 558, 7-21.
(4) S.E. Haagenrud, V. Kucera, J. Gullman. ECS Symposium on Atmospheric Corrosion, October 1980, Hollywood, Florida, USA. pp. 669-693.
(5) T. Sydberger, N.G. Wannerberg. Corrosion Science 12,p 775, 1972.
(6) F.H. Haynie et al. Effect of gaseous pollutants on materials - a chamber study, NTIS PB- 251-580, 1976, 85 pp.
(7) D. Knotkova, B. Bosek, J. Vlckova. In: Corrosion in Natural Environments, ASTM, STP 558, pp. 52-74.
(8) H. Dovland, E. Joranger, A. Semb. Impact of acid precipitation on forest and freshwater ecosystems in Norway. In: F.H. Brække, ed. Ås 1976, (SNSF-project FR 6/76), pp. 14 -35.
(9) F.H. Haynie, J.B. Upham. In: Corrosion in Natural Environments. ASTM, STP 558, pp.31-51, 1975.
(10) D.W. Rice et al., ECS symposium on Atmospheric Corrosion, Hollywood, Florida 1980, pp. 651-68.
(11) O. Svedung, G. Johansson. Proceedings 9th Scandinavian Corrosion Congress, Sept 1983. pp. 337-348.
(12) J.W. Spence, F.H. Haynie. Paint Technology and Air Pollution, Environmental Protection Agency, AT-103, 1972.
(13) H. Dovland, E. Joranger, A. Semb. Wet and dry deposition of pollutants in Sweden, Swedish Water and Air Pollution Research Laboratory, Gothenburg, Sweden 1975.
(14) O.V. Ellis. ASTM Proceedings 49 (1949) pp 152 -170.
(15) S. Haagenrud. Proceedings 8th Scand. Corr. Congr. Aug 1978, pp. 159-69.
(16) OECD. The OECD Programme on Long Range Transport of Air Pollutants - Measurements and Findings. Paris, OECD 1977.
(17) ECS. Symposium on Atmospheric Corrosion. Hollywood, Florida 1980, pp. 873-891.
(18) L. Atteraas et al. Proceedings 8th Scandinavian Corrosion Congress 1978 pp. 139-149.
(19) S. Haagenrud, V. Kucera, L. Atteraas. Proceedings of 9th Scandinavian Corrosion Congress, Copenhagen 1983, p. 257.
(20) B. Knotkova, J. Gullmann, T. Holler, V. Kučera. Assessment of Corrosivity by a Short Term Atmospheric Field Test of Technical Important Metals. Paper for the 6th World Congress on Corrosion in Ottawa 1984.
(21) J.C. Hudson, F.F. Stanners, Journal of appl. chemistry, 1953, p. 86.
(22) L.O. Hagen et al. NILU OR 57/83 Lillestrøm Nov. 83 (in Norwegian).
(23) L.O. Hagen et al. To be published.
(24) ISO/TC 156/WG4 Document N66-69.

STOCHASTIC MODELING OF BUILDING MATERIALS PERFORMANCE IN DURABILITY

A. Siemes, A. Vrouwenvelder and A. van den Beukel

Institute TNO for Building Materials and Building Structures
Delft, The Netherlands

ABSTRACT

The scatter in observed service lifes of buildings and structures is very high. Designing an optimal structure in relation to the durability is therefore not easy. In this paper, the use of a probabilistic approach is explored as a possible means of addressing service life prediction problems. Essentially the same techniques are used as those which have been proven to be successful as a design tool for ultimate and serviceability limit states without deterioration effects. As an example, an outdoor gallery slab, made of reinforced concrete, is analyzed for the limit state of corrosion of the reinforcement. The analysis provides insight into the relative importance of parameters involved. Furthermore, a cost optimisation is performed including building costs and expected maintenance and repair costs.

1 INTRODUCTION

The scatter in service lifes of buildings and structures is, in general, very high. An example of this is given by Bekker [1]. From this study, it follows that the service lifes of houses in the Netherlands varies from about 45 to 125 years. The mean service life is approximately 84 years. The end of the service life is not based in all the cases on technical reasons. Political, economical and social considerations may also have played a part. It is stated, however, that the technical service life of a building also has a high scatter. In general, it is not possible to indicate exactly what the reasons for the scatter are. The reasons, however, are connected with the hardly known and complex set of deterioration mechanisms involved and their interactions. They are also connected with the random nature of the environmental conditions and the structural properties. Even when conditions are apparently the same, the scatter in observed service lifes proves to be very high. In such circumstances, it is not easy for a designer to relate the service life of the various building components to each other and to optimize the design in an economical manner. On many aspects, a great resemblance can be observed between durability on the one side and safety and serviceability on the other side. Among these aspects are:

- the stochastic environmental conditions;
- the stochastic structural properties;
- the mutual relations between building components;
- the economic optimisation of the design.

In the case of serviceability and safety of structures, there is an increasing tendency to approach the design from a probabilistic point of view. Such an approach is called reliability analysis. It seems logical to apply probabilistic methods also in the cases of durability and service life problems. In several publications, such as an approach already has been proposed [2, 3]. The present paper demonstrates that a probabilistic approach is feasible. An additional advantage of a reliability analysis is that it can also be interpreted as a kind of sensitivity analysis. That means that the most relevant parameters in a problem can easily be distinguished, even when only global information is available. This can simplify a problem and serve as a guideline to focus attention on the most relevant aspects. Another advantage originates in harmonising building codes, in which safety, serviceability and durability can be dealt with in the same consistent and logical way.

2 RELIABILITY ANALYSIS

It is necessary to know all the aspects that influence the service life of building materials, elements and structures for quantifying the problem. The first step in a reliability analysis is, therefore, a thorough investigation of all relevant hazards and failure mechanisms. In a systematic way, this can be done by means of a "Failure Mode and Effect Analysis" (FMEA). In a table or checklist a survey is given of all relevant hazards (the causes of deterioration), the belonging failure mechanisms (the way the hazard affects the structure) and the effects (damage). As an example, a FMEA is given in the table 1. This example relates to reinforced concrete. The information for this table is taken from [4-7]. The next step is to distinguish between important and unimportant hazards. Only hazards that might result in effects with unacceptably high risks are taken into further consideration. In this connection, risk is defined as the product of the probability of failure times the amount of damage. If the risks are not known, this means in a probabilistic approach that mathematical models for the failure mechanisms have to be developed. The parameters in these models have to be quantified including their uncertainties.

Table 1: FMEA related to the durability of concrete

	hazards	mechanism	effect
1.	alternating load	fatigue	cracking/failure
2.	streaming water	erosion	surface defects
3.	turbulent water	cavitation	excavations
4.	driving, walking	wastage	impracticability
5.	freezing	expansion	cracking
6.	acids	neutralisation	corrosion
7.	chloride etc.	affecting the passivation layer	pitting corrosion
8.	sulphates	crystal growth	disintegration
9.	de-icing salts	withdrawing heat	scaling
10.	freezing/de-icing	withdrawing heat	scaling
11.	soft water	neutralisation	corrosion
12.	sugar,glycerine	forming of acids	corrosion
13.	micro-organisms	production of acids	corrosion
14.	acidifying gases	neutralisation	corrosion
15.	corrosion	reducing bar diameter	failure
16.	corrosion	rusting	cracking of cover
17.	pollution	crystallisation	pop-outs
18.	alkali aggregate	alkali-silica reaction	expansion

Next it is possible to calculate the probability of failure. Finally the economic consequences have to be determined, with which the risk is known. The simplest mathematical failure model for a building structure consists of one load parameter S and one resistance parameter R, both time independent. Failure will occur when the resistance R is lower than the load S:

$$\{F\} = \{\text{failure}\} = \{R < S\} \qquad \ldots(1)$$

The probability of failure can be calculated from the so-called convolution integral (see figure 1):

$$P\{F\} = \int F_R(s) \cdot f_S(s) \cdot ds \qquad \ldots(2)$$

where $F_R(s)$ is the cumulative distribution function of the resistance R and where $f_S(s)$ is the probability density function of the load S.

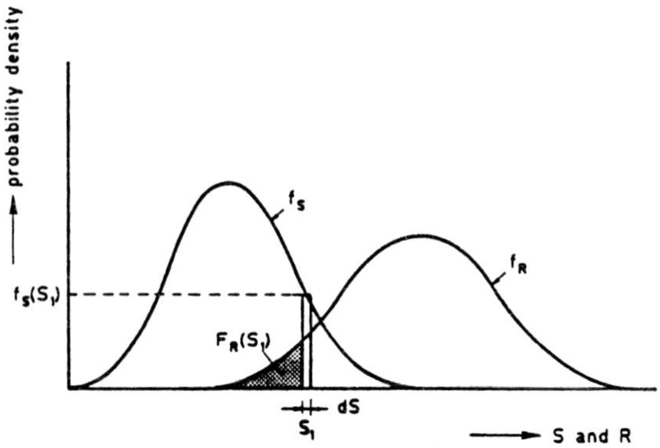

Figure 1: Probability of failure

In the course of time several approximation methods have been developed [3] to work out equation (2). The exact mathematical solution of (2) is in many cases too time consuming to be carried out. In durability problems R and S can be time dependent (see figure 2). Failure occurs if at one moment during the considered period (0 - t) the resistance R(t) is lower than the load S(t) at that moment;

$$P\{F \text{ in } (0-t)\} = P\{R(\tau) < S(\tau) \text{ for at least one } \tau \text{ in } (0-t) \quad ...(3)$$

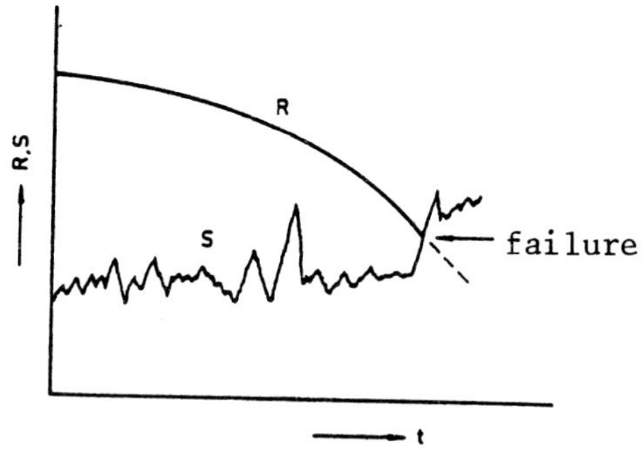

Figure 2: R and S time dependent

Since the event failure in (0 - t) is indentical with the event $\{L < t\}$, equation (3) gives in fact the distribution function F(t) for the service life L (see figure 3a). In this function the failure probability is represented as the area under the curve. The mean μ and the standard deviation σ give an impression of the expected service life and the scatter.
By differentiation once, the probability density function $f_L(t)$ can be obtained (figure 3b). In general calculations on the basis of equation (3) are difficult.

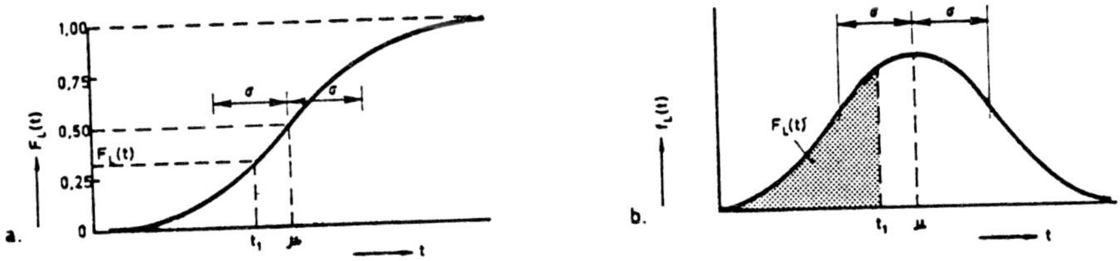

Figure 3: Distribution of the service life L(t)

An alternative procedure is therefore to consider the service life as a sort of resistance L and the time t as a sort of load. Failure in a period (0 - t) can then be formulated as:

$$P\{F \text{ in } (0-t)\} = P\{L < t\} \qquad \ldots(4)$$

This method is also used in the example of the reinforced concrete gallery slab that will be presented further on.

3 ECONOMIC CONSEQUENCES

For an economic optimisation of a design, it is necessary to take into account both the direct building costs and the expected maintenance and repair costs. In fact also energy, administration, cleaning and insurance are costs that should be considered. To simplify the problem, these costs are left out of our consideration. This simplification does not affect the result of the study.

Between the direct building costs and the maintenance and repair costs exist a relation. A right investment can lead to a reduction of maintenance and repair.

In order to establish the optimum (in fact the minimum) costs, it is necessary to compare the direct building costs with the costs that will come in existence in the future. This can be done by converting the expected costs into an amount of money that, in a manner of expressing, has to be reserved next to the direct costs. This reservation is lower than the total amount of the future expenses, because in the course of time it grows due to bearing of interest.
If the target service life of the building equals N years, then the total costs are equal to:

$$E(C_{kap}) = S + \sum_{i=1}^{N} \frac{V_i}{(1+r')^i} + \sum_{i=1}^{N} \frac{P\{F_i\}\, D_i}{(i-r')^i} \qquad \ldots(5)$$

Where:
$E(C_{kap})$ = expected capitalised costs;
S = investment (direct costs);
r' = real interest (nominal interest minus inflation)
$P\{F_i\}$ = probability of failure in year i;
V_i = maintenance costs in year i;
D_i = damage as a consequence of failure in year i;

The advantage of taking into account the real interest is that no allowance has to be made for increasing prices due to inflation.

Instead of capitalising the expected maintenance and repair costs it would also have been possible to make a cost-accounting on annual base. In that case, the investment costs have to be converted to an annual amount for redemption and interest. In principle, the annual and the capitalising method lead to the same conclusions.

4 EXAMPLE: REINFORCED CONCRETE GALLERY SLAB

As a demonstration an outdoor gallery slab (see figure 4) made of reinforced concrete will be optimised. From a FMEA of the gallery slab, it follows that in the Netherlands the main risk is governed by neutralisation of the concrete cover as a result of carbon dioxide. The effect of carbonation will be corrosion of the reinforcement and the consequent failure of the concrete cover (see table 1). All assumptions in this example are as realistic as possible, but are open to discussion.

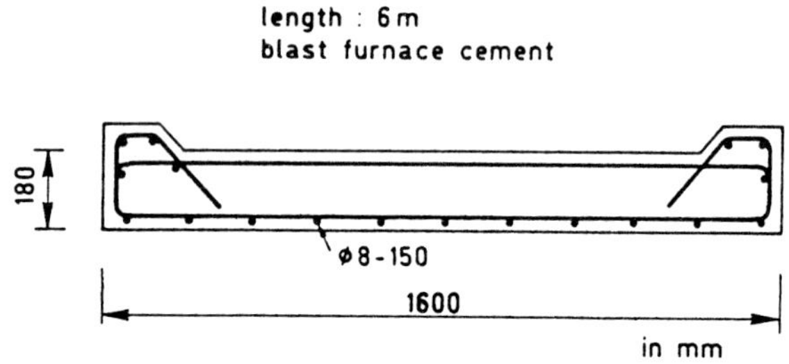

length 6 m
cover 15 mm
w/c ratio 0,50 - 0,55
blag furnace cement
(60% slag)

Figure 4: Reinforced concrete gallery slab

Four design alternatives are distinguished:

1. cover 15 mm, no coating
2. cover 30 mm, no coating
3. cover 15 mm, coating with maintenance every 20 years
4. cover 15 mm, coating with maintenance every 10 years.

The target service life of the slabs is set to 60 years. When damage occurs, there are various repair measures possible, such as replacement of the gallery slab, removing the carbonated zone and placing shotcrete and local repair with synthetic mortar. The kind of repair to be carried out depends upon the damage. In table 2, a cost survey is given for both initial investments, repair and maintenance.

Table 2: Cost survey (in Dutch Guilders per square meter)

new slabs (d=depth)	40 + 0,4.d
replacement for d = 150 mm	275
repair with shotcrete	225
repair with synthetic mortar	250
coating on fresh concrete	50
coating on old concrete	80
maintenance of coating	40
repair of coating	80

From reference [13] a service life relation for the carbonation is derived:

$$L = \left(\frac{(c-\Delta)}{R.K}\left\{\frac{2,7}{46w - 17,6}\right\}\right)^2 + \frac{0,08d}{\phi.v_c} + \frac{(c-\Delta)}{180f_o}\left\{\frac{\frac{T}{T_o}\ln f_o}{1 - e^{e}\frac{T}{T_o}\ln f_o}\right\} \quad ...(6)$$

The first term in this equation describes the actual carbonation process for unprotected concrete [8,9]; the second term represents the time gap until the visibility of the corrosion [10,11]; and the last term is the prolongation of the service life due to the coating [12,13]. In table 3, a survey is given of all the variables in equation (6) and their stochastic properties, viz. the type of the distribution, the mean value and the coefficient of variation (c.o.v). The type of the distribution is denoted with LN for log-normal and D for deterministic.

Table 3: Survey of the carbonation variables

	Description	Type	Mean	C.O.V
c	concrete cover nominal 15 mm	LN	20 mm	0.25
c	concrete cover nominal 30 mm	LN	35 mm	0.14
Δ	distance maximum-mean carbonation	LN	5 mm	0.20
R	cement type parameter	LN	2.0	0.15
K	climate parameter	LN	0.7	0.20
w	water-cement ratio	LN	0.5	0.05
Φ	diameter of reinforcement bar	D	8 mm	-
v_c	corrosion rate	LN	0.04 mm/year	0.50
s	thickness of coating	D	0.18 mm	-
f_o	damage coefficient for coating	LN	0.00001	1.00
T_o	durability parameter	LN	50 year	0.50
T	maintenance period	D	10/20 year	-

The distribution of the service life is calculated on the basis of equation (5) and the parameters in table 3. This is done with the "First Order Second Moment Approximation" method (FOSM) [2,3]. In table 4, the main results, as have been calculated in [13], have been collected. First, the mean service life and the standard deviation have been determined; then the failure probability within a target service life of 60 years is estimated, assuming a log normal distribution for L.

Table 4: Resulting service life distributions (in years)

Design Alternative	Mean Service Life	Standard Deviation	Probability P{L<60 year}
1	34	28	0.76
2	123	86	0.13
3	103	144	0.34
4	417	474	0.05

In figure 5, the service life distributions for the design alternatives 1 and 2 are shown. It is found that the scatter in the service life is very high. In table 5, a survey is given of the contributions of the main variables to this scatter. The contributions result from the FOSM-calculation.

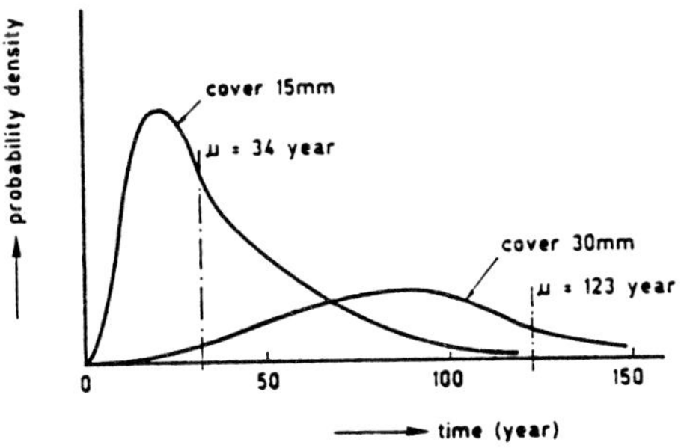

Figure 5: Service life distribution of the designs 1 and 2

The table shows that, as stated in the introduction, the reliability analysis is in fact a sensitivity analysis. In the absence of a coating, the dominating parameters happen to be the cover, the water-cement ratio and the climate parameter. When a coating is applied, the quality of the coating is the most important. Once the service life distributions are known, the expected capitalised costs of each design alternative can be calculated. In table 6, the three terms of equation 4 are presented. They are based on the cost information from table 2 and the service life distributions from table 4. The repair costs are supposed to be dependent on the time of occurence of the damage. The most optimum design proves to be the gallery slab with a concrete cover of 30 mm. It is clear that the application of a coating is even more expensive then alternative 1, despite its long service life.

5 CONCLUSIONS

The optimisation analysis shows that it is possible to use the techniques of the reliability analysis for durability problems.
The reliability analysis provides an estimate of the average life-time and the scatter to be expected.
The variables that are mainly responsible for this scatter can be identified.
An extension of the research into other durability problems seems to be promising.

Table 5: Relative contributions (in %) to the scatter

Description		Design Alternative			
		1	2	3	4
c	cover	52	21	9	10
Δ	distance maximum-each carbonation	2	1	0	0
R	cement type parameter	9	16	0	0
K	climate parameter	17	29	1	0
w	water-cement ratio	19	33	1	0
Φ	diameter of reinforcing bar	–	–	–	–
v_c	corrosion rate	1	0	0	0
s	thickness of coating	–	–	–	–
f_o	damage coefficient for coating	–	–	11	50
T_o	durability parameter	–	–	78	40
T	maintenance period	–	–	–	–
	Total	100	100	100	100

Table 6: Expected capitalised costs (in Dutch guilders)

Cost Term	Design Alternative			
	1	2	3	4
extra concrete cover	–	60	–	–
coating	–	–	500	500
maintenance costs of coating	–	–	860	1250
repair cost expectation	860	170	260	40
$E\{C_{kap}\}$	860	230	1620	1790

REFERENCES

1. Bekker, P. C. F., "Levenscyclus van woningen en de schaarste van grondstoffen," De Ingenieur, No. 39, 1980.
2. Sentler, L., "Service Life Predictions of Concrete Structures," Report TVBK-3018, Lund Institute of Technology.
3. Kraker, A., J. Tichler and A. Vrouwenvelder, "Safety, reliability and service life of structure," Heron, Vol. 27, No. 1, 1982.
4. Leeuwen, J. van and A. Siemes, "Miner's Rule with Respect to Plain Concrete," Heron, Vol. 24, No. 1, 1979.
5. CEB-bulletin 148, "Durability of Concrete Structures - State of the Art Report," February 1982.
6. CUR-VB-rapport 99, "Erosie van beton," Betonvereniging Zoetermeer, Netherlands, 1980.
7. CUR-VB-rapport 64, "Vorstbestandheid van beton," Betonvereniging Zoetermeer, Netherlands, 1974.
8. Nishi, T., "Outline of the studies in Japan regarding the neutralization alkali (or carbonation) of concrete". RILEM, Durability of Concrete, 1970.
9. Kreijger, P. C., "Environment pollution, energy and materials" Materials and Structures/Materieaux et Constructions, Vol. 6, No. 36, 1973, p. 411 - 432.
10. Tuutti, K., "Service life of structures with regard to corrosion of embedded steel". RILEM Congress on Quality Control of Concrete Structures, 1979.
11. Hageman, J. G., "Rapporten, verslagen en notities opgesteld in het kader van de beoordeling van de levensduur van de Oosterschelde". Bureau Hageman, Rijswijk.
12. Klopfer, H., "Die Karbonatation von Sichtbeton und Ihre Bekämpfung B+B," 1 jrg, nr. 3, 1978.
13. Vrouwenvelder, A., A. Siemes and A. van den Beukel, "Duurzaamheid van gebouwen," rapport TNO-IBBC B-83-521/62.3. 1916, October 1983.

VI. FORMULATION OF RECOMMENDATIONS

REPORT OF DISCUSSION GROUP ON INORGANIC MATERIALS

Dr. Julio Ferry Borges
Laboratório Nacional de
Engenharia Civil
Lisbon, Portugal

Participants: Emine Caner
Daniel Dobruse
Göran Fagerlund
J. Ferry Borges, Discussion Leader
Torben Hansen
Gerard Litvan
Hakon Olaffson
Leslie Parrott
Sven Pihlajavaara, Reporter
C. D. Pomeroy
Peter Pratt, Reporter
Th. Tassios

INTRODUCTION

The group developed recommendations for action in the following areas: methodology, modeling as an aid to prediction, real life, materials, and general comments. The recommendations are summarized under headings for each area of discussion.

1 METHODOLOGY

1. The requirements for durability must be defined in terms of quantified performance criteria. From these, the expected service life must be defined either as a lower fractile or as a mean value.

2. Attempts must be made to identify and to understand the mechanisms of deterioration.

3. Guidelines on the general methodologies should be made available to all research workers so that data collection can be done properly so as to assist in modeling.

4. Use must be made of techniques from advanced technologies such as reliability analysis and decision theory.

2 MODELING AS AN AID TO PREDICTION

5. Adequate identification and choice of the basic variables, such as environment, composition, dimensions, time, and temperature is essential. This must be coupled with a probabilistic idealization of these variables.

6. Existing knowledge must be synthesized and expressed in analytical model form for single and for combined deterioration mechanisms. Sensitivity analysis for the basic variables must form part of modeling.

7. Despite the difficulties, the basic variables in the environment must be quantified and synthesized.

8. Accelerated tests should attempt to verify analytical models rather than be used simply to rank materials. The models themselves must be flexible enough to describe both accelerated and in-service tests, leading to improved quality control testing.

3 REAL LIFE

9. Information on durability and on service life of a wide range of materials and components must be collected in a form suitable for interpretation and for modeling. Gross errors of design or of construction should be identified and recorded to avoid their repetition.

10. Part of the gap between materials scientists and engineers can be bridged by regular inspection and control. Quality assurance procedures related to durability must be developed at all phases of concept, design, construction, and use.

4 MATERIALS

11. Concentrated efforts in materials science studies are needed to accelerate the progress to improve the durability of existing materials.

12. New materials and components may have new problems of durability while old materials may change their behavior with time. Studies of these should be included in item 11.

13. Care must be taken with problems arising from interactions between contacting materials in service. Examples include the effect of fire on bond, aggregate-paste interactions, changes in moisture content, and thermal insulation.

5 GENERAL

14. Guidance documents on durability must be improved. Changes made to buildings during their life should be recorded and their influence on service life assessed. Ideally, owners should be given a log book and maintenance manual for their buildings.

15. There is a need for adequate text books on durability for students which include worked examples in service life problems.

16. Life cycle costing, evaluation, and optimization should be made comprehensible and useful to the owner of a building.

REPORT OF DISCUSSION GROUP ON POLYMERIC MATERIALS

Dr. Lars Ostlund
Lund Institute of Technology
Lund, Sweden

Participants: Robert Copé
Philippe Eurin
M. Fiebrich, Reporter
Tore Gjelsvik
Jonathan Martin
Larry Masters
Lars Ostlund, Discussion Leader
J. M. Rigo
Christer Sjöström

INTRODUCTION

The members of the group agreed upon the following headings as discussion topics:

1. Requirements

2. Environmental influences

3. Properties of materials (the materials used in the laboratory as well as the materials in industrial products)

4. Testing (including long term tests, accelerated tests and material control tests) and

5. Methods of service life prediction (including design).

The group also decided to distinguish, in the discussions, the different functional elements comprising buildings and structures, including non-structural elements, protective elements and structural elements. Further, the discussion included new buildings and components, as well as existing buildings and components, and included wooden components, even though wood may not usually be regarded as a polymeric material.

The following is a summary of the results of the discussions according to the headings above.

1 REQUIREMENTS

It was mentioned that a Joint Committee on Structural Safety had formulated requirements, including requirements for durability, in the document "General Principles on Reliability for Structural Design". The requirements are relevant to the discussion of the Workshop in that they are limited to structures and are applicable mainly to wooden structures and polymeric materials used for gluing protective elements or repairing of structural components. There are very few examples of polymeric materials being used as primary structural components.

For components comprised of polymeric materials, a major problem today is that the requirements are imprecise and subjective. There is a need for more precise requirements which, as far as possible, should be formulated in a quantitative way. The subjective nature of the requirements (for example, requirements concerning appearance) could be removed, or at least diminished, by treating them as stochastic variables. In such cases, the determination of quantitative values could be made by a group consisting of a random sample of knowledgable people.

It is important that the requirements provide sufficient flexibility to permit inclusion of economic considerations in material and component selection. Therefore, it is necessary to differentiate the requirements with regard to the intended use of the polymeric material.

An improvement of the situation concerning requirements for polymeric materials and components could be made in two steps. In the first step, all relevant requirements could be identified, differentiated with regard to use, and described qualitatively. Those requirements may, for example, concern color, cracks, delamination of coatings, etc. In the second step, a quantitative description of the requirements could be introduced, taking into account the economic aspects.

2 ENVIRONMENTAL INFLUENCES

In the analysis of deterioration processes of polymeric materials and their components, many difficulties occur. One reason is that there are many kinds of different deterioration processes, which are affected by different types of environmental influences. Another reason is that a deterioration mechanism mainly depends on the micro-climate, which is much more variable than the macro-climate.

There is a great need for a systematic gathering and analysis of climatic data concerning moisture, temperature, ultraviolet radiation, and other climatic factors.

It was mentioned that one possibility might be to let the deterioration of certain well-defined polymeric materials, exposed in a specific way, be a measure of the aggressiveness of the environment. In this way, well characterized polymers may serve as "dosimeters," particularly for solar radiation. Different types of materials would have to be used in order to ensure that results are representative, rather than material specific.

It was stated that it is not sufficient to use only mean values to represent the environmental influences. It is essential to include extreme values as well.

Finally, it was pointed out that, even though it will be difficult and time consuming to obtain the needed quantitative information on environmental influences, steps should be taken to improve the current situation.

3 PROPERTIES OF MATERIALS

It is difficult to identify specific relevant and reliable properties of polymeric materials which should be measured for the generation of information on degradation. This is due to the fact that there are a great number of different types of polymeric materials and each type may have different material properties of importance in assessing its degradation rate or predicting its service life.

With regard to the subject treated at this Workshop, the most important properties of polymeric materials are those essential performance characteristics which are directly affected by deterioration processes. Further, those properties which are of primary interest are not always relevant from an engineering point of view.

When addressing service life prediction problems, it is necessary to have knowledge about the deterioration mechanisms. Thus, research on the basic material properties is essential to develop knowledge about those kinds of polymeric materials which are used for industrial production and about combinations of different polymeric materials. In fact, one of the difficulties in service life prediction is that the link between the properties of "ideal" materials and the properties of building "products" is missing. In the same way, the link between researchers and industrial engineers is missing. An important task of people dealing with service life prediction of polymeric materials is to establish these links.

Also discussed were research needs of polymeric materials which are used for repair of cracked concrete structures by injection, gluing of precast concrete elements and coating of concrete surfaces. For these materials, research is needed on the influence of surface properties, properties of the polymeric materials and application conditions.

4 TESTING

Testing is a very important part of the research on service life prediction. Therefore, it is necessary to develop comparable and reproducible test methods. If this is done, test results from different laboratories can be compared and evaluated on a common basis.

As polymeric materials are often used for repair or protection of, for example, concrete structures, it is important to develop nondestructive test methods and field test methods. Such methods could be used as part of the control of the work concerning protection, repair, or condition assessment.

5 METHODS OF SERVICE LIFE PREDICTION

In order to make it possible to carry out service life predictions with a reasonable degree of reliability, it is necessary to introduce quantitative analyses to a considerably greater extent than that used today. Thus, it is important to reach agreement about general principles which can be used to estimate the service life of a building component or material. These general principles should be a kind of a common framework or methodology for approaching service life prediction of polymeric materials. The methods and approaches used in advanced technologies can, to a great extent, be used to elaborate such a framework of principles.

However, such principles must not function as limits for the development of polymeric materials. It is too early to define standards. If requirements or standards are overly restrictive so that few types of products of polymeric materials are permitted or specified, there is risk that this will hinder development of new technologies.

REPORT OF DISCUSSION GROUP ON METALLIC MATERIALS

Dr. Hans Arup
Korrosionscentralen ATV
Glostrup, Denmark

Participants: Frank Adams
Hans Arup, Discussion Leader
Erik Brandt, Reporter
E. G. C. Coppens
Svein Haagenrud
Christopher L. Page
Steen Rostam
Sam Saunders
Gerhart Schuëller
Ton Vrouwenvelder

1 AREAS OF INTEREST

It was evident from discussions of the previous day that the performance of steel in concrete would attract most of the attention in this final discussion. It should be remembered, however, that there are many other types of constructions using metallic materials for which there is a considerable background of experience with life-time prediction, e.g. 1) painted, galvanized or unprotected steel in open atmospheric exposure, 2) the topic of Haagenrud's paper*, and 3) offshore platforms subjected to corrosion fatigue. A very different type of exposure is presented in technical installations in buildings, e.g. water and effluent pipes, and heating and ventilation systems. These will often have to be updated or replaced during the life of the building, and the associated cost depends heavily

* Included in these Proceedings

on the installation practice. Built-in pipes, for example, can be difficult and costly to replace. After having acknowledged these examples, the group continued to discuss the problems generally, though often with reference to steel in concrete.

The discussion was structured in 3 parts:

a) The need for development and use of tools for service life prediction,
b) The need for data aquisition (field/lab) and
c) Application of results in codes and in teaching.

2 TOOLS

Essential tools in service life prediction are mechanistic data and mathematical models. It is beneficial to use models, even if they are far from final. The models will be improved and revised in the iterative research process. An important use of models is the setting up of a sensitivity analysis, which will help in the selection of worthwhile areas for development and research.

It was said that one should be very hesitant in letting empirical results - or even worse, regression analysis - become the basis for codes of practice, unless the conclusions can find support in some type of model.

Other important tools are all the methods and instruments used for the characterization of materials and environments.

The group strongly supported some statements which were made earlier in the workshop:

- only relevant properties should be measured and
- to have approximate answers to correct questions is better than having precise answers to approximate questions.

Recently developed models for the protection and corrosion of steel in concrete have revealed the need for measurement of "new" properties (which have not been used in the traditional characterization of concrete or cements) in newly formed or installed materials as well as for weathered or aged materials.

Some of these properties have only been measured on laboratory samples, but there is a need for development of methods for "on-site" measurements which, as far as possible, are non-destructive.

The need for the development of non-destructive methods for the measurement of relevant properties on-site exists in very many fields. Permanently placed sensors will allow the monitoring of changes in critical properties with time. There is a need for the development of such sensors, not only to characterize materials, but also to characterize environments, especially the microclimate in closed or partly closed spaces.

3 DATA ACQUISITION

The foregoing discussion naturally leads to the problems of data acquisition. Obviously, the introduction of new kinds of measurements, coupled with the numerous possibilities of automated data collection, could result in unmanageable amounts of data. It has already been emphasized that measurements should be relevant; it could be formulated in this way:

Think first - measure afterwards.

This is equivalent to saying that measurements should not be required unless there is a clear plan for the processing and utilization of results.

The data that are most urgently required in the context of this discussion of service life prediction are such data that can be used as a feedback in the development of models. This means obtaining data for the performance of materials and components in buildings and structures after several years of use, preferably with concurrent monitoring of environmental parameters. It has been mentioned earlier that this has proved difficult in practice for many different reasons (change of ownership, economics).

Some pressure from organizations could help to provide the information required, and some countries have, or have at least considered, codes which require inspections to be made of all buildings, e.g., 5 years after completion. Comparisons were made to the regular inspection of automobiles in Sweden, which has provided substantive and useful feedback - and some pressure - to the car manufacturers. Compulsory inspection of all buildings cannot be expected to give very detailed results, and more thorough reporting about fewer, but typical buildings might give more useful information.

4 IMPLEMENTATION

An improvement in the present situation is only to be expected, if many people agree, that improvements are needed. The most efficient way to influence the development is through codes of practice and this means that one must reach the people sitting on committees.

It is also desirable to introduce the subject in engineering schools and in continuing education.

CONCLUDING REMARKS

Larry W. Masters
National Bureau of Standards
Gaithersburg, MD 20899

INTRODUCTION

The NATO Advanced Research Workshop on Problems in Service Life Prediction of Building and Construction Materials was held to explore opportunities for applying, to research on building and construction materials, analytical procedures which have been utilized in predicting service life of materials, components and systems in advanced technologies. The 32 Workshop participants, representing 16 countries, explored these opportunities in a series of six Workshop sessions; the Workshop Program included invited lectures, small group discussions, reports by group discussion leaders, open discussion periods, question and answer periods with lecturers and a closing summary. The informal interactions and discussion among participants also extended well beyond the Workshop Program.

The purpose of this section of these Proceedings is to present the summary remarks made by the Workshop Chairman at the final session. These remarks were intended to focus upon the key conclusions of the Workshop and to identify the primary recommendations that were made.

THE OPPORTUNITIES

Throughout the Workshop, researchers in building/construction technology expressed recognition of the needs to 1) generate service life data and 2) develop improved tools and procedures for generating these data.

Participants acknowledged that researchers in building/construction technology have not drawn upon analytical procedures to the same extent as researchers in advanced technologies, even though a limited number of researchers in building/construction technology have sought and are seeking to develop scientifically-based prediction methodologies. Thus, the opportunities for researchers in building/construction technology to advance the technology of service life prediction through the use of analytical procedures are vast.

The Workshop clearly pointed out that there are areas of commonality between problems of service life prediction in advanced technologies and those in building/construction technology. Often, for example, the same types of materials are used in both; the need for knowledge on degradation mechanisms is a commonality and frequently the mechanisms may be the same; the need for knowledge of the exposure environment is a commonality and the primary environmental factors are frequently the same even though the intensities may be different; the need for effective characterization tools, particularly non-destructive evaluation methods, is a commonality; the need for reliable accelerated test methods is a commonality, as is the need for reliable predictive models. The fact that commonalities exist between service life prediction problems in advanced technologies and building/construction technology offers promise to researchers in the latter for drawing upon analytical procedures and systematic approaches used by the former.

THE CHALLENGES

Along with the numerous areas of commonality between service life prediction problems in advanced technologies and building/construction technology, Workshop participants also identified numerous differences.

1. Advanced technologies typically utilize simpler materials than those used in building/construction technology and usually incorporate better quality control and quality assurance methods into their processes.

2. The materials and components in advanced technologies may be subjected to less complex and more easily characterized environments than those in building/construction technology.

3. Researchers in advanced technologies frequently utilize fracture mechanics based methods to characterize changing performance and failure but the methods used in building/construction technology are much broader in scope and, frequently less quantifiable, leading to qualitative rather than quantitative assessments.

4. Researchers in advanced technologies frequently define service life expectations and requirements so that they are quantitative and measurable; in building/construction technology, service life has not been defined as effectively. For example, service life is not an absolute number; it is represented by a "failure" distribution, an example of which is shown below.

FAILURE DISTRIBUTION

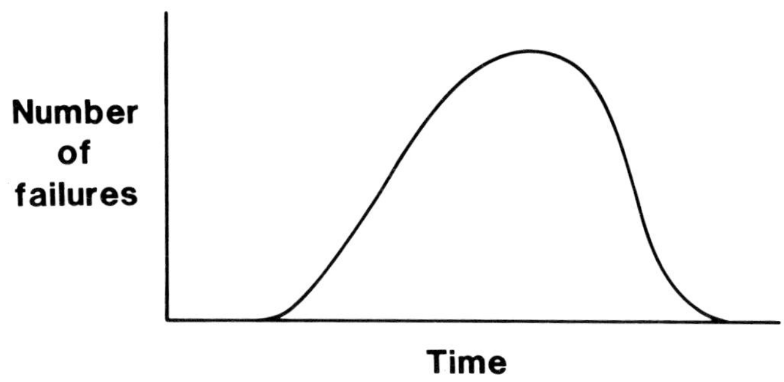

In advanced technologies, service life is typically defined, using probability theory, to indicate an unacceptable number of failures at the low end of the distribution curve. Since this definition reflects the materials or components that fail relatively early (compared to the mean service life), these failures are more easily induced and detected by accelerated testing. In building/construction technology, service life is seldom quantitatively defined; but, in instances in which it is defined, it is usually presented as the mean service life. The failures occurring near the mean service life are more difficult to measure by accelerated testing than early failures, because of the possibility of changing mechanisms of degradation with increasing time along the failure distribution curve and because of the increased time of exposure required to induce these later failures.

5. Researchers in advanced technologies frequently have established feedback loops, providing them with reliable service life data from the field; in building/construction technology, researchers seldom have reliable feedback data.

6. The users of advanced technology systems seem to demand reliable service life predictions, possibly because of the fear of catastrophic failures; in many areas within building/construction technology, there is not the same demand unless life safety is a concern, as for example in the structural elements of bridges and offshore structures.

7. Researchers in advanced technologies rely much more heavily upon accelerated tests for the generation of service life data than do researchers in building/construction technology.

The fact that there are numerous key differences between service life prediction in advanced technologies and building/construction technology presents important challenges to researchers in the latter technology if they are to effectively draw upon procedures and systematic approaches used in advanced technologies.

ACCEPTING THE CHALLENGES

It is clear that a need for improved predictions of service life in building/construction technology exists; it is clear that researchers in building/construction technology have a responsibility to meet the need; it is clear that the analytical procedures used by researchers in advanced technologies can help in responding to the need. But what must be done to address the challenges? The following list summarizes the recommendations made by participants. Where appropriate, quotations captured by the Chairman are noted.

We must develop and utilize a systematic approach to service life prediction so that all essential questions are asked in the research process. Several participants suggested the use of an iterative process, in which hypotheses are made, tested and modified, as needed, until reliable data are forthcoming.

"We need to begin getting approximate answers to the right questions, rather than the right answers to the approximate questions."

We must quantitatively define service life in such a way as to permit its measurement by short-term test methods. Participants suggested that, since service life is a random variable, definition of service life using probability theory is an effective course of action.

"Reliability theory can help in defining the problem - even though little data are available."

We must characterize the environment in which building/construction materials, components and systems are used. This is essential if test methods are to be designed and developed to simulate in-service conditions. Particularly important is the characterization, and perhaps classification, of micro-environments.

We must develop an effective mechanism for obtaining feedback data from the in-service use of materials, components and systems. Prior to the development of such a mechanism, however, it is necessary to define the essential data and how it will be used. Following the development of the feedback mechanism, we must generate representative lifetime data for use in the selection, use and maintenance of materials.

"Though we have limited data on failures, we seldom know why things work well."

We must increase the amount of research on understanding the mechanisms of degradation.

"We have wasted much money in our service life prediction research; what we really must do is to focus our efforts more heavily on understanding degradation mechanisms."

"We need to better understand and account for the synergism between environmental factors that cause degradation."

We must develop improved methods for characterizing materials and components and their degradation processes. Specific needs in characterization include 1) developing fundamentally based data on materials properties and performance, 2) linking microstructure of materials with engineering properties, 3) studying the role of joints between materials in the degradation processes and 4) developing improved nondestructive evaluation tools and methods, particularly for early detection of degradation and for field use (i.e., condition assessment).

"An important stimulus to improved design and improved service life prediction is the development of improved NDE methods which permit the detection of flaws and failures at earlier times."

We must develop improved accelerated or short-term test methods to generate the data needed for service life prediction.

"We need tests that are relevant as well as reliable."

We must do a better job of pointing out the need for improved service life predictions.

"We have to educate our audience to the need, benefits and impact of our research."

"We need to sell the building community on the need for our research."

We must educate ourselves in analytical procedures that could help in service life prediction.

We must develop predictive "models," "hypotheses" or tools for interpreting short-term data in terms of service life.

"Models must be able to account for the past failures and successes as well as be able to predict the future."

"Models must be based on science."

"Reliability theory, as a basis for modeling, can help in linking science and engineering."

NATO ADVANCED RESEARCH WORKSHOP ON PROBLEMS IN SERVICE LIFE PREDICTION OF BUILDING AND CONSTRUCTION MATERIALS

LIST OF PARTICIPANTS

Dr. Frank D. Adams
AFWAL/FIBEC
Wright-Patterson AFB
Dayton, OH 45433 USA

Dr. Hans Arup
Korrosionscentralen ATV
Park Alle 345
Dk 2600 Glostrup
DENMARK

Mr. Erik Brandt
Statens Byggeforsknings Institut
DK 2970 Horsholm
DENMARK

Dr. Emine Caner
Restoration Science Branch
Faculty of Architecture
Middle East Technical University
Ankara
TURKEY

Dr. Robert Copé
Centre Scientifique et Technique du Batiment
24, Rue Joseph Fourier
38400 Saint Martin D'Heres
Grenoble
FRANCE

Mr. E. G. C. Coppens
Ministry of Housing
Boerhavelaan 5
2713 HA Zoetermeer
THE NETHERLANDS

Mr. Daniel Dobruse
Architectural Terra Cotta and Tile, Ltd.
932 W. Washington
Chicago, IL 60606 USA

Mr. Philippe Eurin
Centre Scientifique et Technique du Batiment
24, Rue Joseph Fourier
38400 Saint-Martin D'Heres
Grenoble
FRANCE

Dr. Göran Fagerlund
Cementa AB
P.O. Box 144
S-18212 Danderyd
SWEDEN

Dr. Julio Ferry-Borges
Laboratorio Nacional de Engenharia Civil
Avenida do Brasil, 101
1799 Lisbon Codex
PORTUGAL

Dr. M. Fiebrich
Institut für Bauforschung
Lehr-und Forschungsgebiet
D-5100 Aachen
GERMANY

Mr. Tore Gjelsvik
Norwegian Building Research Institute
Hogskoleringen 7
7034 Trondheim - NTH
NORWAY

Dr. Svein Haagenrud
Norwegian Institute for Air Research
P.O. Box 130
N-2001 Lillestrom
NORWAY

Dr. Torben C. Hansen
Technical University of Denmark
Building 118, SK-2800 Lyngby
DENMARK

Dr. G. G. Litvan
Division of Building Research
National Research Council of Canada
Ottawa KIAOR6
CANADA

Dr. Jonathan W. Martin
Center for Building Technology
National Bureau of Standards
Gaithersburg, MD 20899 USA

Mr. Larry W. Masters
Center for Building Technology
National Bureau of Standards
Gaithersburg, MD 20899 USA

Dr. Hakon Olaffson
The Building Research Institute
Keldnahold
Reykjavik
ICELAND

Dr. Lars Ostlund
Lund Institute of Technology
Fack S-22007
Lund
SWEDEN

Dr. Christopher L. Page
University of Aston
Department of Metallurgy and Materials Engineering
Birmingham B4 7ET
UNITED KINGDOM

Dr. Leslie J. Parrot
Cement and Concrete Association
Research and Development Division
Wexham Springs
Slough SL36PL
UNITED KINGDOM

Dr. Sven Pihlajavaara
Technical Research Centre of Finland
Concrete and Silicate Laboratory
Vuorimiehentie 5
SF - 02150 Espoo 15
FINLAND

Dr. C. Duncan Pomeroy
Cement and Concrete Association
Research and Development Division
Wexham Springs
Slough SL36PL
UNITED KINGDOM

Professor Peter L. Pratt
Imperial College of Science and Technology
London SW 7 2BP
UNITED KINGDOM

Dr. Micheline Regourd
Microstructures Department
CERILH
23, Rue de Cronstadt
75015 Paris
FRANCE

Mst. Dr. Ir. J. M. Rigo
Universite de Liege
6 quai Banning
B-4000 Liege
BELGIUM

Dr. Steen Rostam
COWIconsult
Consulting Engineers and Planners AS
45, Teknikerbyen, Dk - 2830
Virum
DENMARK

Professor Sam Saunders
Department of Pure and Applied Mathematics
348 Sloan Hall
Washington State University
Pullman, WA USA

Dr. Gerhart I. Schuëller
Universität Innsbruck
Institut fur Mechanik
Technikerstrasse 13
A-6020 Innsbruck
AUSTRIA

Mr. Christer Sjöström
Building Planning Division
The National Swedish Institute for Building Research
Box 785
S-801 29 Gävle
SWEDEN

Professor Dr. T. P. Tassios
National Technical University of Athens
42, Patission St.
Athens
GREECE

Dr. Ton Vrouwenvelder
Institute TNO
Building Materials and Building Structures
P.O. Box 49
Delft 2600 AA
THE NETHERLANDS